R00648 92566

S0-DJC-590

REF
QE
511.4
.R43   Recent plate
1987   movements and
         deformation

$20.00

DATE			

BUSINESS/SCIENCE/TECHNOLOGY
DIVISION

© THE BAKER & TAYLOR CO.

Recent Plate Movements and Deformation

Geodynamics Series

Geodynamics Series

1. Dynamics of Plate Interiors
   A. W. Bally, P. L. Bender, T. R. McGetchin, and R. I. Walcott (Editors)

2. Paleoreconstruction of the Continents
   M. W. McElhinny and D. A. Valencio (Editors)

3. Zagros, Hindu Kush, Himalaya, Geodynamic Evolution
   H. K. Gupta and F. M. Delany (Editors)

4. Anelasticity in the Earth
   F. D. Stacey, M. S. Patterson, and A. Nicholas (Editors)

5. Evolution of the Earth
   R. J. O'Connell and W. S. Fyfe (Editors)

6. Dynamics of Passive Margins
   R. A. Scrutton (Editor)

7. Alpine-Mediterranean Geodynamics
   H. Berckhemer and K. Hsü (Editors)

8. Continental and Oceanic Rifts
   G. Pálmason, P. Mohr, K. Burke, R. W. Girdler, R. J. Bridwell, and G. E. Sigvaldason (Editors)

9. Geodynamics of the Eastern Pacific Region, Caribbean and Scotia Arcs
   Ramón Cabré, S. J. (Editor)

10. Profiles of Orogenic Belts
    N. Rast and F. M. Delany (Editors)

11. Geodynamics of the Western Pacific-Indonesian Region
    Thomas W. C. Hilde and Seiya Uyeda (Editors)

12. Plate Reconstruction From Paleozoic Paleomagnetism
    R. Van der Voo, C. R. Scotese, and N. Bonhommet (Editors)

13. Reflection Seismology: A Global Perspective
    Muawia Barazangi and Larry Brown (Editors)

14. Reflection Seismology: The Continental Crust
    Muawia Barazangi and Larry Brown (Editors)

15. Mesozoic and Cenozoic Oceans
    Kenneth J. Hsü (Editor)

16. Composition, Structure and Dynamics of the Lithosphere-Asthenosphere System
    K. Fuchs and C. Froidevaux (Editors)

17. Proterozoic Lithospheric Evolution
    A. Kröner (Editor)

18. Circum-Pacific Orogenic Belts and Evolution of the Pacific Ocean Basin
    J. W. H. Monger and J. Francheteau (Editors)

19. Terrane Accretion and Orogenic Belts
    Evan C. Leitch and Erwin Scheibner (Editors)

Recent Plate Movements and Deformation

Edited by K. Kasahara

Geodynamics Series
Volume 20

American Geophysical Union
Washington, D.C.

Geological Society of America
Boulder, Colorado
1987

 Publication No. 0137 of the International Lithosphere Program

Published under the aegis of AGU Geophysical Monograph Board.

Library of Congress Cataloging-in-Publication Data

Recent plate movements and deformation.

(Geodynamics series ; v. 20) (Publication no. 0137 of the International Lithosphere Program)
1. Plate tectonics. 2. Earth movements. I. Kasahara, K. (Keichi), 1925— . II. Series. III. Series: Publication (International Lithosphere Program) ; no. 0137.
QE511.4.R43 1987   551.1'36   87-14542
ISBN 0-87590-520-X
ISSN 0277-6669

Copyright 1987 by the American Geophysical Union, 2000 Florida Avenue, NW, Washington, DC 20009

Figures, tables, and short excerpts may be reprinted in scientific books and journals if the source is properly cited.

Authorization to photocopy items for internal or personal use, or the internal or personal use of specific clients, is granted by the American Geophysical Union for libraries and other users registered with the Copyright Clearance Center (CCC) Transactional Reporting Service, provided that the base fee of $1.00 per copy plus $0.10 per page is paid directly to CCC, 21 Congress Street, Salem, MA 01970. 0277-6669/87/$01. + .10.
This consent does not extend to other kinds of copying, such as copying for creating new collective works or for resale. The reproduction of multiple copies and the use of full articles or the use of extracts, including figures and tables, for commercial purposes requires permission from AGU.

Printed in the United States of America.

CONTENTS

Foreword   *Raymond A. Price*   ix

Preface   *K. Kasahara*   xi

GROSS MOTION OF THE PLATE SYSTEM AND ROTATION OF THE EARTH

Current Activities in the Measurement of Lithospheric Plate Motion and Deformation   *L. Aardoom*   1

Relations Between the Earth's Rotation and Plate Motion   *Shu-Hua Ye and Koichi Yokoyama*   5

Recent Developments in Space Geodesy: Editorial Note   *Keichi Kasahara*   19

MICRO-PLATES VERSUS PLASTIC DEFORMATION WITHIN CONTINENTAL COLLISION ZONES

Trench Depth and Relative Motion Between Overriding Plates   *Kazuaki Nakamura*   21

Three-Dimensional Numerical Analysis of Continental Marginal Basin Deformation Related to Large Earthquake Development   *Huan-Yen Loo*   27

Neotectonic Deformation of the Alpide Fold Belt in the Central and Eastern Mediterranean and Neighbouring Regions   *N. Pavoni*   35

STRESS AND STRAIN IN THE LITHOSPHERE, IN RELATION TO ANELASTICITY OF THE ASTHENOSPHERE

The Study of Stress and Strain Inhomogeneities at Various Scales in USSR   *G. A. Sobolev*   39

EARTHQUAKES IN RELATION TO PLATE MOTION; ENERGETICS AND DRIVING MECHANISM OF PLATE MOTION

Compilation of Earthquake Fault Plane Solutions in the North Atlantic and Arctic Oceans   *Páll Einarsson*   47

VERTICAL MOVEMENTS AND DEFORMATION OF THE LITHOSPHERE

Recent Crustal Movements in Central Europe   *Pavel Vyskočil*   63

Lithospheric Deformation Deduced From Ancient Shorelines   *P. A. Pirazzoli and D. R. Grant*   67

Recent Crustal Movements and Gravity in Argentina: A Review   *Antonio Introcaso*   73

Summary of the Mid-Term Report, Working Group 1, ICL: Progress in the First Phase of the ILP   *Keichi Kasahara*   81

FOREWORD

Raymond A. Price

Past-President, International Lithosphere Program
and
Director General, Geological Survey of Canada,
601 Booth Street, Ottawa, Ontario, K1A OE8

The International Lithosphere Program was launched in 1981 as a ten-year project of interdisciplinary research in the solid earth sciences. It is a natural outgrowth of the Geodynamics Program of the 1970's, and of its predecessor, the Upper Mantle Project. The Program — "Dynamics and Evolution of the Lithosphere: The Framework of Earth Resources and the Reduction of Hazards" — is concerned primarily with the current state, origin and development of the lithosphere, with special attention to the continents and their margins. One special goal of the program is the strengthening of interactions between basic research and the applications of geology, geophysics, geochemistry and geodesy to mineral and energy resource exploration and development, to the mitigation of geological hazards, and to protection of the environment; another special goal is the strengthening of the earth sciences and their effective application in developing countries.

An Inter-Union Commission on the Lithosphere (ICL) established in September 1980, by the International Council of Scientific Unions (ICSU), at the request of the International Union of Geodesy and Geophysics (IUGG) and the International Union of Geological Sciences (IUGS), is responsible for the overall planning, organization and management of the program. The ICL consists of a seven-member Bureau (appointed by the two unions), the leaders of the scientific Working Groups and Coordinating Committees, which implement the international program, the Secretaries-General of ICSU, IUGG and IUGS, and liaison representatives of other interested unions or ICSU scientific committees. National and regional programs are a fundamental part of the International Lithosphere Program and the Chairman of the Coordinating Committee of National Representatives is a member of the ICL.

The Secretariat of the Commission was established in Washington with support from the United States, the National Academy of Sciences, NASA, and the U.S. Geodynamics Committee.

The International Scientific Program initially was based on nine International Working Groups.

Copyright 1987 by the American Geophysical Union.

WG-1 Recent Plate Movements and Deformation
WG-2 Phanerozoic Plate Motions and Orogenesis
WG-3 Proterozoic Lithospheric Evolution
WG-4 The Archean Lithosphere
WG-5 Intraplate Phenomena
WG-6 Evolution and Nature of the Oceanic Lithosphere
WG-7 Paleoenvironmental Evolution of the Oceans and Atmosphere
WG-8 Subduction, Collision, and Accretion
WG-9 Process and Properties in the Earth that Govern Lithospheric Evolution

Eight Committees shared responsibility for coordination among the Working Groups and between them and the special goals and regional groups that are of fundamental concern to the project.

CC-1 Environmental Geology and Geophysics
CC-2 Mineral and Energy Resources
CC-3 Geosciences Within Developing Countries
CC-4 Evolution of Magmatic and Metamorphic Processes
CC-5 Structure and Composition of the Lithosphere and Asthenosphere
CC-6 Continental Drilling
CC-7 Data Centers and Data Exchange
CC-8 National Representatives

Both the Bureau and the Commission meet annually, generally in association with one of the sponsoring unions or one of their constituent associations. Financial support for scientific symposia and Commission meetings has been provided by ICSU, IUGG, IUGS, and UNESCO. The constitution of the ICL requires that membership of the Bureau, Commission, Working Groups, and Coordinating Committees change progressively during the life of the project, and that the International Lithosphere Program undergo a mid-term review in 1985. As a result of this review there has been some consolidation and reorganization of the program. The reorganized program is based on six International Working Groups:

WG-1 Recent Plate Movements and Deformation
WG-2 The Nature and Evolution of the Continental Lithosphere
WG-3 Intraplate Phenomena

WG-4 Nature and Evolution of the Oceanic Lithosphere
WG-5 Paleoenvironmental Evolution of the Oceans and the Atmosphere
WG-6 Structure, Physical Properties, Composition and Dynamics of the Lithosphere-Asthenosphere System

and six Coordinating Committees:

CC-1 Environmental Geology and Geophysics
CC-2 Mineral and Energy Resources
CC-3 Geosciences Within Developing Countries
CC-4 Continental Drilling
CC-5 Data Centers and Data Exchanges
CC-6 National Representatives
    Sub-Committee 1 - Himalayan Region
    Sub-Committee 2 - Arctic Region

This volume is one of a series of progress reports published to mark the completion of the first five years of the International Geodynamics Project. It is based on a symposium held in Moscow on the occasion of the 26th International Geological Congress.

Further information on the International Lithosphere Program and activities of the Commission, Working Groups and Coordinating Committees is available in a series of reports through the Secretariat and available from the President — Prof. K. Fuchs, Geophysical Institute, University of Karlsruhe, Hertzstrasse 16, D–7500 Karlsruhe, Federal Republic of Germany; or the Secretary-General — Prof. Dr. H.J. Zwart, State University Utrecht, Institute of Earth Sciences, P.O. Box 80.021, 3508 TA Utrecht, The Netherlands.

R.A. Price, President
Inter-Union Commission on the Lithosphere, 1981-85

PREFACE

The motion of the lithospheric plates is the most fundamental aspect of plate tectonic theory. Measurement of their on-going motion and their internal deformations will certainly verify and elucidate this important theory. The technological progress in the geosciences during the last decade stimulated the international geoscience community to address this fundamental problem when Working Group 1 (WG1) of the International Lithosphere Program (ILP) was established in 1981.

As with other ILP Working Groups, a principal challenge of WG1 has been to promote the necessary international coordination of various research groups sharing common interests. In this context, the objectives of our group were: to measure the motion and deformation of the plates, both contemporary and during the recent geological past; to establish the distribution of stress and strain in the lithosphere and to develop models for the way these are concentrated; and to provide the basic data required for modelling of the internal mechanisms responsible for driving the plates. Polar motion and variations in the earth's rotation were studied to determine possible relationships between these quantities and geophysical phenomena such as earthquakes and volcanic eruptions. A wide variety of geophysical, geodetic, and geological data were needed for these purposes.

The Working Group was established in 1981 with 16 regular members and 11 corresponding members. The number of corresponding members was increased to 25 by the end of 1985, to foster cooperation with other interested groups, namely: the Commission on Recent Crustal Movements (CRCM), the International Association for Quaternary Research (INQUA), and its Commissions on Shorelines and Neotectonics; and International Geological Correlation Programme (IGCP) Projects Nos. 146, 200, 201 and 202. International meetings co-sponsored with these groups included four major symposia (Tokyo, 1982; Hamburg, 1983; Wellington, 1984; Moscow, 1984), in which WG1 played the lead role.

This volume is a report of progress consisting of a summary of the past activities of the Working Group, together with 10 other papers that are divided among five topics: (1) gross motion of the plate system; (2) micro-plates versus plastic deformations within continental collision zones; (3) stress and strain in the lithosphere, in relation to anelasticity of the asthenosphere; (4) earthquakes in relation to plate motion, energetics and driving mechanism of plate motion; and (5) vertical movements and deformation of the lithosphere. This grouping of papers was chosen so that the reader may correlate the individual papers to the WG1's objectives, and may have an overview of the progress and future problems in each research field. The conclusions and recommendations of the Working Group outline the general direction of research in the field in the next half decade.

There has been remarkable progress in the study of recent plate motion and deformation during the past five years. The targets assigned to the Working Group, which initially looked remote and blurred, now appear close and clear because we have moved forward. It is my cordial hope that the International Lithosphere Program will continue to make progress during the remainder of the 1980's and that the final report of Working Group 1 will record the successful completion of the main objectives.

K. Kasahara
Volume Editor

Copyright 1987 by the American Geophysical Union.

CURRENT ACTIVITIES IN THE MEASUREMENT OF LITHOSPHERIC PLATE MOTION AND DEFORMATION

L. Aardoom

Department of Geodesy, Delft University of Technology
P.O. Box 5030, 1600 GA Delft, The Netherlands

Abstract. An assessment of the gross motion of the plate system is most efficiently obtained by the application of accurate space geodetic positioning techniques. Techniques currently applied are laser ranging to artificial satellites and very long baseline interferometry using stellar radio sources. Their potential accuracies over distances up to several thousands of kilometers render them viable candidates for the monitoring of plate motion by repeated relative positioning. For effectiveness monitoring program should strive for international coordination and global extent. Exsiting programs are reviewed as to their objectives, methods and results obtained. A preliminary assessment of ongoing plate motion in selected areas is expected before 1990, but considerable progress, both in instrumentation and international coordination, will be required in order to obtain a reliable global picture.

Introduction

Available kinematic models of global lithospheric plate motion (e.g. Minister and Jordan, 1978) are based on combined interpretation of a variety of geological and geophysical evidence. As a consequence such models describe a pattern of motion averaged in some way over the interval of time to which the analysed data refer. Therefore the models may not necessarily be representative of current or present crustal motion. Geodetic determination of station displacements, however, has the advantage of yielding contemporary quasi-instantaneous motion values. In general, the more accurate these measurements are, the shorter the time span of the measurements can be. Geodetic methods thus may provide a picture of contemporary, ongoing, motion relevant to the understanding of destructive phenomena. Although repeated geodetic relative position determinations lead to direct estimates of actual surface motion, such surface motions are no more than a set of boundary conditions, albeit important ones, for unraveling the mechanism of plate tectonics.

The monitoring of the gross motion of the lithospheric plate system, including both the quasi-rigid motion of the major plates and their broad scale deformation, is a task well suited to modern space geodetic positioning

Copyright 1987 by the American Geopysical Union.

techniques. In addition to high precision the important features of space techniques are:
(a) a global distribution of repeated station positions;
(b) the three-dimensionality of position determinations.

Only the most precise of the space geodetic techniques are candidates for monitoring lithospheric motion in an efficient way. For mapping the gross motion pattern only those of the precise space techniques capable of operation over intersite distances of over several hundreds or thousands of kilometers are to be considered: laser ranging to artificial satellites (satellite laser ranging: SLR) and independent-clock stellar radio-interferometry (very long baseline interferometry: VLBI). Other space geodetic positioning techniques, such as satellite-based radio-interferometry (e.g. the U.S. Global Positioning System: GPS), have as yet not demonstrated satisfactory precision over the distance range considered, but may definitely be used to regionally densify the monitoring network. Difficulties of implementing the technique of laser ranging to moon-based retroreflectors at a sufficient number of globally distributed sites have prevented this technique from directly contributing and will probably continue to do so. This leaves two potential techniques for the purpose: SLR and VLBI. Together with satellite-based radio-interferometry (GPS) terrestrial geodetic and related methods may be applied to provide the regional and local details of the deformation in selected areas of interest. Among these methods are high precision horizontal surveying, precise levelling, strain gauging and gravimetry.

Current geodetic programs of monitoring the gross motion of the lithospheric plate system are based on the application of either SLR or VLBI or both of these techniques. The development of these techniques will be reported on elsewhere in this issue.

Current Programs

The problem of monitoring the gross motion of the plate system is essentially one of global scope. There are several reasons, however, for programs to initially focus on selected features of the overall pattern of motion or on selected geographic areas. These are:
(a) the available accuracy of station position determinations and the time span required to produce

reliable results tend to restrict monitoring to networks where large strains are expected;

(b) available instrumentation for SLR and/or VLBI is mostly sited at permanent stations, which restricts its involvement to the monitorintg of part of the inter-plate motion pattern only;

(c) the limited availability of transportable SLR- and/or VLBI-equipment calls for concentrated efforts in a limited number of selected areas of scientific priority.

Without pretending to be exhaustive the following is meant to be a representative summary of individual programs, running or being in an advanced stage of preparation.

The Crustal Dynamics Project initiated by the U.S. National Aeronautics and Space Aministration (NASA) was established in 1979 as a merger of a number of existing national projects (Flinn, 1981). Although the Crustal Dynamics Project is basically a national one, NASA seeks and succeeds in finding cooperation with other countries in order to widen the scope towards a virtually global one. The efforts to monitor plate motion at large exploit permanent and transportable SLR- and VLBI-systems operated under the project or cooperatively by other countries. More regionally the project presently concentrates on areas in the Western United States, Western Canada and Alaska and in the Eastern Mediterranean. The latter area is the subject of a joint venture together with European investigators.

A consortium of European investigators is dedicated to the monitoring of crustal motion in the Mediterranean area (WEGENER, 1984). For the time being their main interest is focused on the Eastern section of the Mediterranean where, in close cooperation with the countries there, a program of repeated SLR is being developed using both permanent and transportable equipment. The program is executed jointly with NASA in the framework of the Crustal Dynamics Project.

European radio-telescopes have begun to cooperate as a coordinated VLBI facility with precise relative positioning and earth kinematics as program elements. Several of these radio-telescopes take part in bilateral programs, e.g. in the NASA Crustal Dynamics Project.

In the framework of the Intercosmos Programme the Astronomical Council of the Academy of Sciences of the U.S.S.R. is developing a project of regional and broad scale geodynamics employing permanent SLR equipment in Eastern Europe, Asia and South America. The inclusion of VLBI is being studied (S.K. Tatevian, private communication, 1984).

In Japan a number of scientific institutes contribute to a national program, in which the monitoring of plate motion and deformation by means of space geodetic techniques is an important element. Both SLR and VLBI are employed (K. Kasahara, private communication, 1984). Although the program has the investigations of regional tectonics in the area as its primary objective, bilateral cooperative agreements ensure a contribution to programs of more global extent.

Australia being located on the supposedly fairly rigid Indo-Australian plate, is considered a good platform for evaluating the temporal consistency of the precise space geodetic techniques and for monitoring plate motion with respect to or in the continent's surroundings (Stolz and Lambeck, 1983). These topics are the primary elements of a developing program based mainly on U.S.-deployed and Australian SLR equipment. Transportable SLR is being considered. Geodetic VLBI, incorporating both fixed and transportable equipment, is also being discussed for implementation in the program. There is close cooperation with NASA, presently in the context of the Crustal Dynamics Projects. A joint activity with New Zealand is under discussion.

Program Characteristics and Results

Of the quoted national, extended national or multinational programs of space geodetic monitoring of the gross motion of the plate system, only NASA's Crustal Dynamics Project is in the actual execution phase. Due to a relatively large number of bilateral agreements this projects has moreover the widest geographical scope. Yet it is not a truly international one.

An example of a truly international project, and consequently basically a global one, is Project MERIT (to Monitor Earth-Rotation and Intercompare the Techniques of observation and analysis), initiated in 1978 by the IAU and cosponsored by the IUGG since 1979 (Wilkins, 1980). Project MERIT concerns the rotation of the earth and necessarily of its crust, subdivided as it is into tectonic plates on which the observing stations (applying SLR, VLBI or other techniques) reside. The project in its present form is limited to a 14-month observing period, however, which practically excludes inferences about relative motion and deformation of the plates. To operationally define earth-rotation, a conventional terrestrial reference system has to be established as part of the project and this will intrinsically serve as the zero-epoch frame to which global plate motion is referred.

Mostly in the framework of the Crustal Dynamics Project, a rather impressive number of interstation chord lines has been determined in recent years, crossing the boundaries of several of the major lithospheric plates. Such determinations are based on either SLR or VLBI. Repeat determinations enable a first assessment of present-day rates of relative plate motion. Similar repeated determinations of chord lines within plate enable an initial assessment of plate deformation. Results of these assessments are to be reported elsewhere in this issue. Those based on SLR, for example, do not appear to be inconsistent with available kinematic models (Christodoulidis et al., 1985). This is somewhat surprising because, as indicated in the Introduction, available models necessarily reflect motion averaged over the past few million years; it might be expected that the present pattern of motion would substantially deviate from the average one. If the deviation is indeed minor, the geodetic assessment will have yielded an indication that away from their boundaries the major lithospheric plates move at constant relative rates.

The above results suggest that a sufficient number of space geodetic monitoring sites should be chosen away from deforming boundary zones, so that relative station motion may be interpreted in terms of plate motion or broad scale deformation of plate interiors. Site selection and maintenance should be accompanied by repeated precise and reliable local or regional ground surveys in order that apparent site instability can be detected and taken into account. Attention should be paid to unambiguous reoccupation of sites consistent with the precisions of repeated relative positioning.

Concluding Remarks

Modern geodetic positioning techniques and space geodetic techniques in particular appear to provide unique data on the global pattern of plate motions.

As a result of their inherent precision, both SLR and VLBI are suitable space geodetic techniques. Because of their differing sensitivities to uncontrolled sources of error, these techniques should be regarded as complementary, rather than competing. Both techniques need further improvement, not only in precision, but also in ease of handling, transportability and cost of operation.

Such improvement together with indispensible developments in both the theory of analysis and alternative instrumental approaches, are likely to lead to considerable progress in the assessment of the gross motion pattern by 1990. To obtain a reliable global picture, however, a dedicated global network configuration should be realized. To achieve that, a more intensive coordination of individual programs will be required. It is expected that the development of individual programs will gradually bring about the required cooperation.

References

Christodoulidis, D.C., D.E. Smith, R. Kolenkiewics, S.M. Klosko, M.H. Torrence and P.J. Dunn, Observing tectonic plate motions and deformations from satellite laser ranging, J. Geophys. Res., 90, 9249-9263, 1985.

Flinn, E.A., Application of space technology to geodynamics, Science, 213, 89-96, 1981.

Minster, J.B. and T.H. Jordan, Present day plate motions, J. Geophys. Res., 83, 5331-5354, 1978.

Stolz, A. and K. Lambeck, Geodetic monitoring of tectonic deformation in the Australian region, J. Geol. Soc. Australia, 30, 411-422, 1983.

Wegener, several papers, Ann. Geophys., 2, No. 2 and 3, 1984.

Wilkins, G.A. (ed.), Project MERIT, a review of the techniques to be used during Project MERIT to monitor the rotation of the Earth, Publ. by Royal Greenwich Observatory, England and Institut fur Angewandte Geodasie, F.R.G., 1980.

RELATIONS BETWEEN THE EARTH'S ROTATION AND PLATE MOTION

Shu-hua Ye

Shanghai Observatory, Academia Sinica, People's Republic of China

Koichi Yokoyama

International Polar Motion Service, Mizusawa, Iwate, 023 Japan

Abstract. Relations between the Earth's rotation and plate motion are reviewed from various points of view to foresee geodynamical progresses which will be achieved by new space techniques. First, astronomical and geophysical evidences of the Earth's variable rotation and plate motion are presented. Then geophysical relations between perturbations in the Earth's rotation and various dynamical processes relating to plate motion are presented. Also described are observational relations between the Earth's rotation and displacements of the points on the Earth's surface, which are subject to global plate motion. Finally given are fundamental requirements for terrestrial and celestial reference systems, in which the Earth's rotation and plate motion are described.

I. The Motion of the Earth

The Earth, on which we live, is in continuous motion of various types.

As a planet, the Earth is revolving around the barycenter of the solar system. This orbital motion can be described as a motion of a point mass under the gravitational forces of the Sun, Moon, and the planets.

On the other hand, the Earth is rotating around its center of mass. The Earth's rotation is not uniform, but is perturbed by various parts of the Earth mutually interacting, such as the core, mantle, crust, ocean, and atmosphere. Perturbations in the Earth's rotation can be conceptually classified into three aspects, namely, precession and nutation, polar motion, and variation of the length of day (lod).

In addition, geophysical processes occurring inside the Earth induce global plate motions, and regional and local crustal deformations.

From the viewpoint of geodynamics the Earth's rotation and plate motion are two major phenomena which relate to each other both geophysically and observationally. Hence in order to understand global dynamics, which governs the Earth's rotation and plate motion, precise monitoring of the perturbations in the Earth's rotation, and inter- and intra-plate motions is absolutely necessary.

Recently achieved dramatic improvement of precision and accuracy in radio interferometry and laser ranging has enabled us to measure these motions with centimeter accuracy, which may be enhanced to millimeter accuracy in the next decade [Walter, 1984].

1. The Earth's Rotation

a. Precession and Nutation

Precession and nutation, which are the forced motions of the Earth's spin axis due to the luni-solar torques, directly reflect the internal constitution of the Earth. It is presently difficult to determine the precession constant theoretically with sufficient accuracy, because the internal constitution of the Earth is not well modeled for that purpose. The precession constant adopted in the IAU 1976 System of Astronomical Constants was deduced from the proper motions of stars in our galaxy with the accuracy of about ±0.1"/century [Fricke, 1971]. Very long baseline interferometry (VLBI), which observes extra-galactic radio objects, has feasibility to determine the precession constant free of galactic rotation, with an accuracy of a few milliseconds of arc per century. More precise determination of the precession constant contributes to constructing a more precise nutation table.

The formerly adopted nutation table [Woolard, 1953] was constructed on the basis of the rigid Earth model which never reflects the internal constitution of the real Earth. As early as 1902, defect of the rigid Earth nutation theory, which was in use at that time, was revealed as the z-term in the equation of observation for estimating the polar motion [Kimura, 1902]. Since then, a number of theoretical and observational work

Copyright 1987 by the American Geophysical Union.

have been made to determine the nutation of the real Earth. It turned out that the nutation of the rigid Earth is remarkably different from that of the real Earth, especially in the 18.6-year, annual, semi-annual, and fortnightly terms.

Hence the IAU and IUGG decided to adopt the theory by Wahr [1981], which uses the Earth model by Gilbert and Dziewonski [1975]. The new theory of nutation assumes an inertial core-mantle coupling. If one assumes, however, a viscous coupling, the amplitude and phase of the 18.6-year component may be slightly modified [Sasao et al., 1977, 1979]. Ocean loading can also yield amplitude modification. Confirmation of the phase and amplitude modifications, which are expected to be achieved by new techniques, will remarkably contribute to the progress of Earth dynamics.

Free-core nutation, whose period may be nearly one sidereal day, is another important phenomenon to be confirmed. This resonance effect due to dynamical ellipticity of the fluid outer core has not been confirmed from observations by classical techniques, because of its small and time-varying amplitude. When the time-variation of the amplitude of the free-core nutation is detected by the new techniques, knowledge of resonant and damping mechanisms of the core-mantle boundary will be extremely increased.

It is noteworthy that the solid inner core may produce a free nutation of another period.

b. Variation of the lod

b.1. Secular variation. Paleontological evidences show that the lod has become longer with an averaged rate of 2 ms/century [Wells, 1966]. Analyses of historical records of astronomical observations of the Sun, Mercury, eclipses, and occultations give an averaged rate of 1-2 ms/century [Spencer-Jones, 1939; Stephanson and Morrison, 1983]. Based on various estimations from ancient astronomical records, Lambeck [1980] concluded 2.07 ±0.19 ms/century as the rate of astronomically observed rate of the lod variation. On the other hand, tidal deceleration estimated from tidal theory, and astronomical and satellite observations with ocean tides being taken into account gives 2.67 ms/century. Difference of the two values may indicate secular acceleration of the lod. If one introduces, however, an assumption of variation of the gravitational constant, the non-tidal acceleration becomes insignificant [Lambeck, 1980]. It is considered that the secular rate of the lod have been constant over the last 3000 years.

b.2. Decade fluctuations. Decade fluctuations of the lod have characteristic time scales ranging from some five years to a few decades with the amplitudes of about a few milliseconds. Due to the lack of long term data from space techniques, investigation of the decade fluctuations has relied upon optical astrometry observations from the early 19th century.

Explanations of the decade fluctuations have been sought in the mechanism of the core-mantle coupling. Electromagnetic coupling, rather than viscous coupling, may be important. Topographic coupling, which assumes irregularity of the core-mantle boundary, is also possible [Hide, 1982]. Detailed review was made by Rochester [1974] on the various coupling mechanisms and astronomical consequences. The decade fluctuations could, in part, be due to the change of the Earth's moment of inertia associated with small sea level variations. Sources of the decade fluctuations were searched for in the activity of sunspots, and the motion of the Moon and the planets [Luo et al., 1974; Zheng and Zhao, 1979].

b.3 Seasonal variations. The strongest components in the seasonal variations are the annual, semi-annual, and biennial ones, whose amplitudes are about 0.4, 0.3, and 0.1 ms, respectively. Two other components also exist, whose periods are about a few months and shorter than a month, and their UT1 amplitudes are 1-2 ms and about 0.5 ms, respectively [Guinot, 1974; Zheng, 1978]. The semi-annual variation is mainly induced by the change of the inertial moment of the solid Earth due to the zonal tides. On the other hand, other seasonal variations are induced by the change of atmospheric angular momentum due to zonal winds and redistribution of atmospheric mass. This implies that the seasonal variations of the lod can never be stable in amplitudes. Particularly, the periodicity of the biennial component varies from year to year [p.e. Langley et al., 1981].

Carter et al. [1984] showed that the UT1 series obtained during three years from 1980 both by VLBI and SLR agree very well with each other. VLBI and laser ranging are promising techniques to detect fine structures of irregular variations of the lod. Expected accuracy of the lod in the next decade is ±0.02 ms/day [Walter, 1984].

b.4. Atmospheric angular momentum. Rosen and Salstein [1981] evaluated the lod change based on the data compiled at the National Meteorological Center (NMC) of the USA. Currently daily values of the atmospheric lod based on the NMC data are regularly evaluated and distributed in the MERIT (an international project by The IAU/IUGG Joint Working Group on Earth Rotation to Monitor Earth Rotation and Intercompare Techniques of observation and analysis) group. In addition to the NMC, both the European Center for Medium-Range Weather Forecasts and the Japan Meteorological Agency provide the global analysis data, which form the basic data for evaluating the atmospheric lod. The astronomical and atmospheric lod's agree with each other within the rms error of ±0.2 ms. By improving the accuracy of the atmospheric lod, it is now possible to investigate the relation between El Niño Pacific warming and sudden changes of the lod [p.e. Rosen et al., 1984]. This kind of task is extremely important to investigate dynamics of the solid Earth, by which the decade fluctuations of the lod are probably governed.

c. Polar Motion

c.1. Chandler wobble.
The Chandler wobble is the only one actually confirmed component among the theoretically predicted free nutations.

As early as 1765, Euler predicted an existence of 305-day free nutation. In 1891, Chandler found a 14-month component in the astronomical latitude variation, but no 305-day component. Immediately after the discovery of the 14-month component, Newcomb showed that elasticity of the Earth and fluidity of ocean lengthen the period of the free nutation from ten to fourteen months, and hence the 14-month component detected by Chandler is a free wobble of the real Earth. Thus this term was named Chandler wobble.

Theoretical estimations have shown that the Chandler period is lengthened by the elastic mantle and ocean, and shortened by the liquid core. For example, the Chandler period is estimated as 434.9 days for the model with a neutrally stratified core inside the elastic mantle whose Q value is assumed to be 300, and with a refined ocean equilibrium theory [Smith, 1977; Dahlen, 1976]. A non-equilibrium ocean theory changes the estimation only by 0.2 days. The frequency dependent Q model, however, changes the Chandler period considerably. On the other hand, the Chandler period determined from observations is 434.4 days [Jeffreys, 1968].

It has been suggested that the wobble amplitude fluctuates from 0.05" to 0.25" with a period of about 40 years. The fluctuation of the amplitude seem to be interrelated to that of the Chandler period. This may be possible by the non-equilibrium pole tides.

Since the Chandler wobble is a free nutation, it should be maintained by some excitation mechanisms. Sources have been searched for in seismic and aseismic activities, and atmospheric excitation. Numerous controversy took place on seismic excitations of the Chandler wobble. Main problem is in the evaluation of seismic moment which is given by moment-magnitude relations. The seismic effect on the polar motion has not yet been confirmed observationally, mainly due to the noise level of optical astrometry observations, which is comparable to the polar shift expected theoretically. On the other hand, atmospheric excitation is now believed to be a major source which maintains the Chandler wobble. Routine computations of the atmospheric excitation functions of the Chandler wobble, as well as the atmospheric lod, are in progress at the three organizations (see b.4).

The Chandler wobble will decay, when no energy is supplied, because of the anelasticity of the Earth, which is generally described using the quality factor Q as

$$1/Q = 1/(2\pi E) \int \Delta E \, dt$$

where E is the maximum energy stored during a period and ΔE is the energy dissipated. No systematic decay of the amplitude, however, has been found in the astronomical observations.

c.2. Seasonal fluctuations.
The annual forced polar motion, whose amplitude is about 0.1," is caused by seasonal redistribution of atmospheric and oceanic mass, and ground water. On the other hand, an existence of a forced semi-annual fluctuation has not been confirmed yet from optical astrometry observations. Even though it exists, its amplitude will not exceed 0.005".

c.3. Secular polar motion.
The eighty-year ILS (International Latitude Service) polar motion shows a secular drift of the pole superimposed by long periodic oscillations [Markowitz, 1970]. The rate and direction of the secular drift, so far derived, are about 0.003"/yr and 70° West longitude.

Reality of the long periodic oscillations of the pole has been a matter of debate. The point has been, whether they are due to the errors in the adopted star catalogs or not. A homogeneous series of the ILS polar motion was computed by Yumi and Yokoyama [1980], in which star positions and proper motions were compiled in a uniform system. This series of the polar motion also shows the existence of the long periodic oscillations [Okamoto and Kikuchi, 1983]. This implies that the long periodic oscillation found in the ILS polar motion is not primarily due to the star catalog errors. Zhao and Zheng [1980] found periodic components ranging from a few years to tens of years in the polar drift. Several common periods were also identified in the atmospheric excitation functions. This may suggest that the atmosphere, coupled with the oceans, may play some role to induce a long periodic polar motion.

2. Plate Motion

a. Internal Structure of the Earth

The Earth is known to consist of the crust, mantle, outer core, and inner core. There exist three discontinuities between them: Moho (3-80 km), Gutenberg (2900 km), and Lehmann (5200 km) discontinuities. The crust is the region above the Moho discontinuity, which varies in depth below the Earth's surface from 3 to 10 km beneath the oceans, and from 55 to 80 km beneath the continents.

The upper mantle extends from the Moho discontinuity to about 400 km including both the lower part of the lithosphere and the upper part of the asthenosphere. The lithosphere (50-150 km) is considered to be a shell with high strength overlying the asthenosphere with relatively low strength. The asthenosphere extends from the bottom of the lithosphere to about 700 km in depth, and is weak in the sense that it deforms by creep. The transition zone is about 600 km thick, where the body wave velocity is considerably reduced.

The lower mantle extends from 1000 km to the Gutenberg discontinuity. The lower part of the transition zone and the lower mantle are collectively referred to as mesosphere.

In the outer core, S wave does not propagate. This means that the outer core consists of liquid. S wave can propagate in the inner core, although the velocity is slow. This suggests that the inner core is composed of molten or partially molten materials.

b. Plate Tectonics

Morgan [1968] introduced the concept of plate tectonics. That is, the Earth's surface is composed of a mosaic of the lithospheric plates that exhibit some combinations of: oceanic rises or divergent plate boundaries, subduction zones or convergent plate boundaries, transform faults, and intra-continental compressional zones. According to plate tectonics, movements of the continents and sea-floor are considered to be a part of large-scale movements of plates. The plates range from about 10^4 km^2 to 10^8 km^2 in size, and from about 70 km beneath the oceanic region to 150 km beneath the continental region in depth. Plate boundaries do not in general coincide with continental margins.

Contemporarily there are seven major plates: the Eurasian, Antarctic, North American, South American, Pacific and Australian plates. There are some intermediate or small plates as well. Plates may diminish or grow in size, depending on the distribution of convergent and divergent boundaries.

Hess [1962] first proposed the concept of sea-floor spreading; the blocks of sea floor are moving relative to one another in response to convective currents within the mantle. This is the process by which the oceanic lithosphere splits at oceanic rises and moves away from rise axes. At oceanic rises the plate separates and the lithosphere is generated. Transform faults are created by strike-slip displacement at right angles to the mid-ocean ridges. Oceanic lithosphere is consumed in the asthenosphere at subduction zones to create new lithosphere.

Evidences supporting the concept of the sea-floor spreading have come from various sources, such as the magnetic stripes of the ocean floor, the age of ocean volcanic rocks, and the age of deep-sea sediments. Rates of sea-floor spreading (expressed as half-rates) deduced from ocean magnetic anomaly patterns range from 1 to 20 cm/yr; on the average a few centimeters per year.

In 1915 Wegener proposed a hypothesis that the present continents were once comprised in a large super-continent which he named Pangaea. According to Wegener, Pangaea began to break apart and individual continents started moving toward their present positions during the Mesozoic Era. The idea of continental drift, however, did not receive wide acceptance for a long time. The breakthrough came, when the sea-floor spreading theory became widely approved. The sea-floor spreading theory offered a means to drift the continents and to retain their deep roots on the base of the lithosphere, not on the Moho discontinuity.

c. Characteristics of Plate Motion

The continental drift is the movement of the continents relative to one another across the Earth's surface, as a result of sea-floor spreading which brings about plate motions. In the case of rigid plates constrained to move on the surface of a sphere, the motions will be only rotations [Le Pichon et al., 1973]. Hence the motion of one plate relative to another along the mutual interface can be described in terms of the motion of an axis of rotation passing through the center of mass of the Earth. Relative plate motions are best measured by the angular velocities. The axis of rotation of a plate relative to another is referred to as the pole of spreading. The velocity increases from zero at the pole to a maximum at the equator of rotation.

Poles of spreading and spreading rates, as well as sizes and shapes of the plates, change with time. The sizes and shapes generally change gradually, while the spreading centers often change abruptly over the period as short as 10^5-10^6 years, as deduced from magnetic anomaly patterns. Movements of small plates ($\leq 10^6$ km^2) appear to be controlled by compressive forces of larger plates.

d. Driving Mechanism of Plates

A viscous-drag model assumes a strong coupling between lithosphere and asthenosphere, such that the upward convection beneath oceanic ridges and lateral spreading drag plates [Richter, 1973].

The plates may move away from the ocean ridge crests, since the gravity component of the lithospheric plates on the flanks of the oceanic rises pushes the plates away [Hales, 1969].

Gravity sliding is another possible driving mechanism. The colder lithosphere, chemically the same as the underlying hotter asthenosphere, is reasonably be considered to be denser. Thus the lithospheric plate edge sinks due to gravity sliding, and is supposed to pull the rest of the plate toward the trench and away from the ridge. In this theory the mantle plays a passive role.

The buoyancy model assumes an asthenosphere that is soft enough not to transmit significant horizontal shear stresses to the overlying lithosphere. Plate movement is caused by a combination of: pushing apart of plates at oceanic ridges, gravity sliding of plates away from ridges, and pulling of plates by their leading edges as they descend at subduction zones [Elsasser, 1971].

II. Geophysical Relations Between the Earth's Rotation and Plate Motion

1. Irregular Rotation of the Earth in the Geological Past as a Possible Cause of the Break-up of the Super-Continent

A major question concerning plate tectonics is, what is the primary cause which brought about the break-up of the super-continent. Although the boundaries of the existing major continents are irregular, it is possible to find the boundaries which fit each other. In this fitting one would notice that two kinds of lines can be drawn along the boundaries. One is more or less parallel to the equator, while the other is oblique, making similar angles with the equator on both sides of the boundaries. What kind of mechanism made such boundaries? Goldreich and Toomre [1969] proposed a hypothesis that a large angular displacement of the Earth's pole occurred in geological time scale owing to gradual redistribution of density irregularities within the Earth by the process of mantle convection. Such a shift of the polar axis with respect to the mantle must cause stresses and deformation of the thin shell, since it would adjust its shape to the change of the flattening. Hence the stresses and deformations generated by the polar shift may provide a mechanism for the original break-up of the shell into several major tectonic plates. Turcotte and Oxburgh [1973], and Freeth [1980] put forward the basic concept to membrane tectonics. They demonstrated that under certain circumstances, the generated stresses can be large enough to induce lithospheric fractures. Liu [1974] calculated the state of stress existing in the shell and used the plastic theory to analyze the possible break-up of the shell. His result gave a pattern similar to the current boundary system of the major plates as described by Le Pichon [1968] and Isacks et al. [1968]. Song [1979], using a more compact and efficient method, concluded that with the increase of the polar shift (several tens of degrees) in geological time scale, the existing boundaries of plates were gradually formed.

2. Earthquakes as a Possible Source of Excitation of the Chandler Wobble

The Chandler wobble has not decayed since the beginning of modern astronomical observations, although it is a free nutation subject to damping. Manshinha and Smylie [1967], and Smylie and Manshinha [1968] reopened the study of excitation mechanism of the Chandler wobble by earthquakes, which was initiated in the late 19th century. The proposed mechanism is such that the displacement fields induced by earthquakes cause the change of the products of inertia, by which the Chandler wobble is excited. This relates to two questions. One is the development of the dislocation theory to describe the deformation of the Earth due to the surface loading, and the other is the estimation of seismic moment induced by an individual earthquake, which is essential in deciding whether earthquakes are sufficient sources of excitation to maintain Chandler wobble or not. An individual earthquake causes a shift in the pole path.

Magnitude and direction of the polar shift due to the 1960 Chilian and 1964 Alaskan earthquakes were estimated by Dahlen [1973], Israel et al. [1973], O'Connell and Dziewonski [1976], Smith [1977], and Manshinha et al. [1979]. They give similar results both in magnitude and direction, in spite of conceptual differences in the treatment of the liquid core and core-mantle boundary conditions. The estimated polar shifts are about 0.02" and 0.01" for the respective earthquakes. Similar estimation by Song et al. [1981], based on a modified liquid core equation, gives the polar shift of about three times larger than the above values, although they used the same fault parameters as Dahlen [1973] did. Slade et al. [1984] studied the post-seismic effects of great earthquakes at plate boundaries on polar shift, using the 3-dimensional finite element technique for a variety of asthenospheric rheology and low viscosity zones.

Smylie and Manshinha [1968] searched for the expected seismic effects in the BIH polar motion, and found a strong correlation between the times of occurrence of major earthquakes and changes in the curvature of the pole path. Haubrich [1970], however, questioned that the breaks in the pole path identified by Smylie and Manshinha [1968] reflect observational noises. The new space techniques will provide sufficiently accurate data to confirm the predicted polar shift. It is also expected to identify pre- and post-seismic effects in the pole path by the new techniques.

A cumulative effect of seismic activities has been investigated extensively. Myerson [1970], for example, found a strong correlation between amplitude variation of the Chandler wobble and the number of earthquakes occurred in a year, but he concluded that earthquakes themselves are not primary sources of excitation. Correlation with the released seismic energy was also suggested [Anderson, 1974]. Smylie and Manshinha [1971] and Dahlen [1971] pointed out the importance of seismic activities in the excitation of the Chandler wobble. Dahlen [1973], however, reviewed his earlier results, and concluded that earthquakes can not excite the Chandler wobble, mainly because of over-estimation of the seismic moment estimated by seismic moment-magnitude relations. O'Connell and Dziewonski [1976], based on the 30 largest earthquakes during 1901-1964, gave a Chandler excitation of 0.1" which is close to the observed level, 0.15".

Thus seismic excitation of the Chandler wobble

is still an open question. Observational confirmation is in the realm of the new techniques.

Hide [1984] suggested that the atmospheric excitation could account for the observed polar motion, without finding it necessary to invoke the Earth's core or effects due to movements in the solid Earth associated with earthquakes of magnitude not exceeding 7.9.

It is most probable that the excitation of the Chandler wobble is a combined effect of seismic activities and atmospheric mass shift.

3. The Secular Polar Drift and Plate Motion

Proverbio and Quesada [1974] and Poma and Proverbio [1980] emphasized a possibility that the observed secular polar motion of the ILS is due to displacements of lithospheric plates where ILS stations are located. Lambeck [1980] showed that the observed polar drift, 0.003"/yr, is too large to be explained by station displacements inferred from the plate motion model by Minster and Jordan [1978]. It is, however, suspicious whether plate motions inferred from paleomagnetism reflect contemporary station displacements at plate margins, where ILS stations are located. Okamoto and Kikuchi [1983] suggested that Ukiah is mainly responsible for the long periodic oscillations in the polar motion. Since Ukiah lies near San Andreas Fault, large episodic motions may have occurred during these eighty years.

Another possibility is the change of the products of inertia in geological time scale. Due to sea-floor spreading, oceanic lithosphere splits at oceanic ridges and is consumed at subduction zones to create new lithosphere. This mass shift, if not entirely consumed at subduction zones, may change the products of inertia. The rate of seafloor spreading is comparable to that of the observed secular polar motion.

Accumulation of the polar shifts by individual earthquakes is another possibility [Manshinha et al., 1979]. Identification of the breaks in the pole path will give us an answer on this effect.

4. The Period and Q Value of the Chandler Wobble and the Inner Part of the Earth

The Chandler period, estimated theoretically using a realistic Earth model and a refined ocean theory, agree with the observed one as closely as a few days. Anelasticity of the mantle represented by Q value changes the estimation of the Chandler period by: 1.8 days for Q=300, 3.9 days for Q=600, and 8 days in the case of frequency dependent Q. Thus the Chandler period is a good indication of the mantle anelasticity. Importance of the Chandler wobble in the estimation of Q value lies in the fact that its period (about 430 days) is far longer than both seismic (1 second - 1 hour) and tidal (0.5 - 1 day) ones. Dissipation mechanism of the Chandler wobble is reasonably considered to be different from that of seismic and tidal frequency bands. In combination with seismic and tidal Q values, Chandler Q affords invaluable information on evaluating frequency dependence of the Q value [Anderson and Minster, 1979]. Further, the coefficient in the Jeffreys' [1972] modified Lomnitz [1962] law may be derived [Smith and Dahlen, 1981; Okubo, 1982]. Thus precise evaluation of the Chandler period and Q value is essential in understanding dissipation mechanism of the mantle.

The Chandler Q also plays an important role in evaluating atmospheric excitation of the Chandler wobble. Observed Chandler Q values lie between 60 and 100, and it is now believed that the Chandler Q is closer to 100. If this is true, excitation of the Chandler wobble can be explained by atmospheric mass shift to a considerable extent.

On the other hand, phase shifts of the diurnal tides are also invaluable sources of information on the mechanism of core-mantle coupling, which in turn is closely related to excitation of the free-core nutation, and phase shift of the 18.6-year nutation. By analyzing strain observations, Sato [1985] derived 700 days as the damping time of the free-core nutation, which corresponds to Q=1500. This gives an upper limit of the kinematic viscosity of the core of some 10^5 cm^2 s^{-1}. Astronomical confirmation of the free-core nutation and the phase shift of the 18.6-year nutation are of major importance in geodynamics. Core viscosity is closely related to turbulence in the liquid core flow, which in turn affects the Chandler period and mechanism of maintaining geomagnetic dynamo [Li and Song, 1981; Hinderer et al., 1982].

III. The Observational Relations Between the Earth's Rotation and Plate Motion

1. General Principle

Various techniques have been developed to make regular monitoring of the Earth's rotation since the 19th century. Recently new techniques, such as VLBI, satellite laser ranging (SLR), and lunar laser ranging (LLR), succeeded in achieving more than two orders of magnitude improvement of precision, when compared with visual telescopes.

In spite of differences in observational techniques, the basic principle of observation of the Earth's rotation is the same. That is, each technique determines the station vector with respect to the source vector defined by celestial objects or artificial satellites, whose positions and motions are described in the inertial reference system. In other words, what is determined is a "relation function", which relates the source vector to the station vector at the instant of observation.

The definitions of the station and source vectors, and the relation function are:

- Source vector **O**: A vector from the Earth's center of mass to an observed object. For distant objects, such as stars in our galaxy and extra-galactic radio objects, it is regarded as a unit vector from any point of the Earth.

- Station vector **S**: A specially designed vector through the observation site.
- Relation function F: A function which describes the relation between **O** and **S**. Its expression depends on geophysical models adopted and is known. Its instantaneous parameters are measurable.

Let O_I and S_I denote the source and the station vector referring to the inertial reference system, and O_E and S_E the source and the station vector referring to an Earth-fixed reference system.

We define:
- M_r : The rotation matrix to transform the true equatorial system of date to the inertial system.
- M_t : The rotation matrix to transform the inertial system to the true equatorial system of date.
- N_r : The rotation matrix to transform an Earth-fixed system to the true equatorial system of date.
- N_t : The rotation matrix to transform the true equatorial system of date to an Earth-fixed system.

Then we get the relations:

$$S_I = M_r N_r S_E$$

and

$$O_E = N_t M_t O_I .$$

Given the expression and instantaneous values of the relation function, we get

$$F(O_E, S_E) = F(N_t M_t O_I, S_E)$$

or

$$F(O_I, S_I) = F(O_I, M_r N_r S_E) .$$

Then we get an equation of condition, which contains the parameters related to: the source vector O_I, the station vector S_E, the Earth's rotation, station displacements, Earth tides, positions of celestial objects, astronomical and geodetic constants, and so on.

We can select various combinations of the unknown parameters depending on the given conditions.

Examples:
a. Given the vector O_E (from the ephemeris), the matrix M_t (from the astronomical constants), and the vector S_E (from the adopted station coordinates), one can get the elements of the matrix N_t, namely, the Earth rotation parameters (ERP's: precession and nutation, polar motion, and lod or UT1), using the data of multiple stations.
b. Given the quantities S_E, as well as M_t and N_t, or M_r and N_r, the source vector O_I can be derived from multiple observations of the same source.
c. Given the quantities O_I, as well as M_t and N_t, or M_r and N_r, the station vector S_E can be derived.

Thus the ERP's and station positions are included in a relation function in an unseparable manner.

2. Expressions of the Rotation Matrices

M_t or M_r, which rotates the coordinate system by the precession and nutation, can be written as [Lieske, 1979]

$$M_t = P(-\epsilon_A - \Delta\epsilon) R(-\Delta\psi) P(\epsilon_A) R(-Z_A) Q(\theta_A) R(-\zeta_A)$$

and

$$M_r = M_t^{-1}$$

where matrices $P(\alpha)$, $Q(\alpha)$, and $R(\alpha)$ represent the rotation about the x, y and z axes, respectively. $\Delta\psi$ and $\Delta\epsilon$ are the nutation in longitude and in obliquity, ϵ_A the mean obliquity, and ζ_A, Z_A and θ_A are the precession angles. All these quantities are taken from the IAU 1976 System of Astronomical Constants and the IAU 1980 Theory of Nutation. Readers may refer to the MERIT Standards [Melbourne et al., 1983].

N_t or N_r, which rotates the coordinate system by the polar motion and UT1, can be written as

$$N_t = Q(-x) P(-y) R(+u) R(+\theta_E)$$

and

$$N_r = N_t^{-1}$$

where x and y are the instantaneous coordinates of the pole in the left-hand system with reference to the internationally adopted origin, u is UT1-UTC, and θ_E is the ephemeris mean sidereal time. u is the perturbation in the speed of the Earth's rotation measured with reference to the internationally coordinated uniform clock, UTC. $R(+\theta_E)$ means that the Earth-fixed coordinate system rotates uniformly. The rotation angle defined in the IAU 1976 System of Astronomical Constants is calculated, based on the reading of the atomic time as the argument.

Ignoring $R(+\theta_E)$ for simplicity, we get

$$N_t = \begin{pmatrix} 1 & u & x \\ -u & 1 & -y \\ -x & y & 1 \end{pmatrix}$$

and

$$N_r = N_t^{-1}$$

where x, y and u are small quantities.

Thus the expressions of the vectors **O** and **S** can be derived as the functions of the ERP's.

3. The Definition of the Relation Function

a. Optical Astrometry

The station vector **S** is the vector along the local vertical, and the source vector **O** is in the direction of a star from the Earth. Both **S** and **O** are unit vectors. Any type of instrument measures a relative angle between **O** and **S**. When the zenith distance z of a star is measured to determine instantaneous latitude, the relation function is

$$F(O_E, S_E) = \cos z = O_E \cdot S_E.$$

When we measure hour angle t to derive UTO-UTC, the relation function is

$$F(O_E, S_E) = \cos t = (O_E(x) + O_E(y)) \cdot (S_E(x) + S_E(y))$$

where x and y denote the x and the y component of a vector.

b. VLBI

The source vector is the vector to an extragalactic radio object from the Earth, while the station vector is the baseline between two stations. Observables are the delay and delay rate of the wave front between two stations. The relation functions, in this case, are

$$F(O_E, S_E) = c\tau = O_E \cdot S_E$$

and

$$F(O_E, S_E) = c\dot{\tau} = \frac{d}{dt}(O_E \cdot S_E)$$

where c is the light velocity.

c. Laser Ranging

Observed objects are artificial satellites and the Moon, which are both in the limited distances from the Earth. The source vector, in this case, is from the mass center of the Earth to that of a satellite or the Moon. Motion of a satellite or the Moon is given by the dynamical theory. What is measured is the distance ρ between an observation site on the Earth and a reflector on a satellite or the Moon. Provided that the vector **O** is corrected for the libration or the attitude variation, so that it is referred to the mass center, the relation function is

$$F(O_E, S_E) = \rho^2 = (O_E - S_E) \cdot (O_E - S_E).$$

d. Doppler Tracking

Laser ranging measures the distance, while Doppler tracking measures the difference of the distance. Then the relation function is

$$F(O_E, S_E) = \Delta((O_E - S_E) \cdot (O_E - S_E))^{\frac{1}{2}}.$$

The most essential difference between classical optical astrometry and the new techniques is that the former measures angular distances, while the latter measures light-time. Effect of atmospheric refraction is far larger in the measured angular distance than in the light-time. This is one of the the major reasons why optical astrometry is poor in precision.

4. Separation of Earth Rotation from Plate Motion

Motion of a station vector, with reference to the inertial system, contains both the Earth's rotation and "proper" motion of the station. They can not be separately estimated, unless an appropriate constraint is introduced.

Generally, we need to set up a terrestrial reference system fixed on the surface of the deformable Earth in some averaged sense, which can be realized by a set of stations. The rotational motion of this system with respect to the inertial system is regarded as the Earth's rotation, if a constraint is taken in such a way that there is no translation nor rotation in the terrestrial reference system due to the individual station proper motions. The residual motions then give station proper motions defined in an appropriately chosen terrestrial reference system.

5. Present Situation of Each Technique

The classical technique started monitoring the polar motion in the late 19th century, and the variation of the lod in 1955 when precise atomic clocks were introduced for time keeping. Currently about one hundred instruments are in operation to contribute to the BIH (Bureau International de l'Heure) and the IPMS (International Polar Motion Service) work to provide the ERP's of optical astrometry. Precisions of the BIH 5-day ERP's are about ±0.01" for polar motion and ±1 ms for UT1. On the other hand, smoothed daily EOP's of the IPMS are provided with formal errors of about ±0.003" and ±0.2 ms, respectively.

The method of satellite Doppler tracking has provided the polar motion, since 1969, with the precision comparable to that of optical astrometry. Recently this technique achieved considerable improvement of precision by tracking the drag-free NOVA satellite.

The network of satellite laser ranging is composed of about 30 stations, half of which are equipped with the second or the third generation instruments whose precisions are ±10-20 cm and a few centimeters, respectively. The laser ranging technique can provide the pole coordinates and UT1 with precisions of about ±0.002" and ±0.3 ms, respectively [Wilkins, 1984].

Since 1983, the VLBI, even though its network is composed of only a few regular and some additional cooperative stations, has regularly provided, not only the pole coordinates and UT1, but also two components of the nutation, with precisions even better than ±0.001" and ±0.1 ms. The IRIS (International Radio Interferometric Surveying), an IAG subcommission, has published IRIS Earth Orientation Bulletin [p.e. IRIS, 1984].

Close agreement between the atmospheric [p.e. Rosen et al., 1984] and the astronomical lod derived from the new techniques, introduced a new prospect into the future of international Earth rotation service and geophysical researches.

During the main campaign of Project MERIT, the BIH published the weighted mean values the ERP's, based on the data provided by all available techniques of observation. Relative weights given to the classical techniques were only 3% for polar motion and 30% for UT1, while those given to the new techniques were 97% and 70%, respectively. Although it is questionable whether or not such a simple weighting system is reasonable to determine the ERP's, the above relative weights tell us the present accuracy of each technique.

The VLBI and laser ranging are two promising techniques capable of measuring plate motions in several years, as accurately as a few millimeters per year. Experiments are under way in southern California based on these techniques, to measure relative motion between the Pacific and the North American plate, and 6 cm/yr was obtained as a preliminary result [Musman, 1982]. Other projects to measure global, regional, and local displacements are also in progress involved in the NASA's Geodynamics Project. Detailed review of the project was made by Walter [1984].

IV. Reference Systems for Determining the Earth's Rotation and Plate Motion

1. Basic Requirements for the Terrestrial System

Dynamics of the Earth is governed by geophysical processes of various time scales. Mantle convection and tectonic motions are the phenomena of the lowest frequency band, while seismic and tidal waves belong to the highest frequency band. On the other hand, the perturbations in the Earth's rotation lie between the two extreme frequency bands.

On the basis of spatial extents of geophysical processes, the following classification may be useful: i) global phenomena like the Earth's rotation and global plate motion, ii) regional phenomena which contain inter- and intra-plate motions, and iii) local phenomena due to tectonic strain accumulation along the plate boundaries.

Different types of phenomena can be described using different types of reference systems. For example, the hot-spot reference system is useful for describing paleomagnetic and geological phenomena which belong to the lowest frequency band [Morgan, 1972; Stefanick and Jurdy, 1984; Jurdy and Gordon, 1984].

In this paper, which mainly deals with global phenomena of medium time scale measured with geodetic tools, we confine ourselves to considering the geodetic reference system, in which global deformations and perturbations in the Earth's rotation are determined.

a. Dynamical Requirements for the Terrestrial System

A reference system should be simple and reliable. Invariant phenomena should remain invariant when expressed in that system. Geodetic parameters deduced referring to such a system should be interpreted meaningfully, in terms of relevant geodetic and geophysical effects. In particular the system should be maintained on permanent basis on the deformable Earth in some averaged way, in order that observations made at different epochs can be brought together to deduce useful information of the Earth's rotation and deformations.

From dynamical viewpoint, the Tisserand axes are most preferable to deal with the Earth's rotation and global plate motions [Munk and McDonald, 1960]. From the definition of the Tisserand axes [Moritz, 1980], the relative angular momentum

$$\mathbf{h} = \int \mathbf{x} \wedge \mathbf{u} \, dV$$

due to motion $\mathbf{u}$ relative to the $\mathbf{x}$-system should be zero. Munk and McDonald [1960] called the axes defined by the integration over the whole Earth "Tisserand's mean axes of body".

Approximately we can regard these axes as being "fixed" in the deformable Earth in some averaged way. It is not very useful, however, to make integration over the whole Earth, since the core and the mantle possibly undergo different rotation. A more preferable choice is to define the axes with respect to, either the crust and mantle or only the crust [Smith, 1981]. Dynamically the former choice may be better. Actually, however, we need to know the orientation of a point on the crust where an instrument resides. The axes defined by integrating over the crust only is to be called "Tisserand's mean axes of the crust". This definition suits geodetic purposes, since geodetic observations are made on the crust. For dynamical treatments of the mantle, some difficulty might be introduced.

b. Kinematical Requirements for the Reference System

Although we made conceptual definition of the reference system from the dynamical viewpoint in the preceding section, practical realizations are possible only by kinematical and geometrical approaches.

A polyhedron may be one of the realizations of the terrestrial reference system. What is essential is to form a highly accurate and dense network, which is composed of the stations distributed along the plate boundaries and also on the plates. The 3-dimensional positions of these stations, which compose the vertices of the polyhedron, are measured at regular and frequent time intervals. The axes of the reference or the 3-dimensional coordinates of the vertices can be chosen arbitrarily at an initial epoch. Usually they are selected, so that the system may have continuity to the BIH-CIO axes currently in use. At a later epoch the shape and the orientation of the polyhedron will change, due to various deformations and motions. The new shape and orientation of the polyhedron are deduced in terms of the three dimensional coordinates of the vertices, with respect to the reference system which is "fixed" in the deformable Earth.

The question is how to determine the new coordinates. Geometrically only the change of the polyhedron shape could be determined. We can not estimate the coordinates and the ERP's simultaneously, since the determinant of the normal equation becomes zero. Hence additional conditions or constraints need to be introduced. Mathematically various constraints are possible to be introduced. For example, one can introduce inner-constraints or pseudo-inverse method to solve a rank deficient equation. The "minimum principle" method [Richter, 1981] is often used. This method requires that the square sum of all deformations or velocities be minimum.

Kinematically the system should be defined in such a way that there is neither common rotation nor translation in the variations of the coordinates of the stations [Mueller, 1981; Bender and Goad, 1979]. In spite of the variety of possible choices of the constraints, all these constraints are basically equivalent to one another [Bock, 1982].

Let us consider whether the reference system defined by these constraints fullfils dynamical requirements. Zhu [1983] showed that if an infinite number of stations are uniformly distributed over the whole crust, the reference system defined under the above constraints is essentially equivalent to the Tisserand's mean axes of the crust. Strictly speaking, the axes which are arbitrarily selected at the initial epoch usually do not coincide with the Tisserand's axes. The above mathematical procedure proves that the axes of the selected system keep fixed relations with the Tisserand's axes, no matter how the Earth deforms itself.

The function of the system is twofold. One is to monitor motion of the terrestrial system, or the perturbations of the Earth's rotation, with respect to the inertial system to be realized by the VLBI using extra-galactic objects. This is a motion of the polyhedron as a whole. The other is to monitor relative internal motions of the polyhedron, or combined effect of tectonic motions and other sources of deformation including vertical motions. These two kinds of motion, which are subject to a constraint adopted, can be estimated using an appropriate technique of analysis.

c. Practical Problems

c.1. Modeling deformations. For establishing and maintaining the terrestrial reference system, various deformation models are to be incorporated. Tidal deformations of the solid Earth, including ocean loadings, are well modeled. The plate motion model by Minster and Jordan [1978], may it be the best available, is not certain whether it really reflects contemporary rates and directions of the plate motions. Introduction of an improper model will degrade the quality of the estimates. Models will be refined by observations, and even new models will be required in an unexpected way to explain observations. Refined models will reduce residuals of observation.

c.2. Network. Currently neither the number of dedicated stations nor their geographical distribution is adequate for realizing the terrestrial reference system, which is considered as a sufficient approximation of the Tisserand's mean axes. Setting up of at least a few stations on each of the major plates, and frequent visits by mobile VLBI, SLR, and GPS are required.

C.3. Colocation. VLBI and laser ranging are indispensable tools to establish and maintain the future terrestrial reference system. The geophysical implication of the terrestrial reference system is, however, different from technique to technique. The former is based on the geometrical reference and the latter on the dynamical reference. Their roles in establishing the terrestrial reference system are complementary. Connection of the geometrical and dynamical reference systems will lead to establishing a unique terrestrial reference system for use by all techniques.

As the first step to implement such a unique terrestrial reference system, colocated measurements were proposed by the COTES (The Joint IAG/IAU Working Group on the Conventional Terrestrial System) [Mueller et al., 1982]. The observational campaign was carried out during the main campaign of Project MERIT. One of the objectives of COTES activities is to clarify the relations among various reference systems. In order to realize this, systematic errors and errors in physical models were carefully examined in the course of the MERIT main campaign [Wilkins, 1984].

c.4. Other problems. In order to evaluate a combined series of the ERP's using the data provided by various techniques, the BIH has adopted relative weights. This procedure is worthy of being reviewed. Formal uncertainty, which is often adopted for establishing a weighting system, does not necessarily reflect external accuracy. In addition, geophysical meanings of the ERP's depend on the observational principle of each technique. Hence a more sophisticated mathematical technique may be required in the near future, to estimate the ERP's as the geophysical parameters common to all techniques of observation.

It often happens that stations participate in or drop out of the network. This causes change of polyhedron configuration, and makes it difficult to preserve the initial system.

2. Celestial Reference System

The celestial reference system realized by a set of astronomical constants directly affects the determination of the Earth's rotation, baseline length, and station coordinates.

Each technique rests upon its own celestial system, whose origin and orientation differ from technique to technique. Hence the origin and orientation of the terrestrial reference system, as well as the ERP's, are affected by the celestial system adopted [Zhu and Mueller, 1983].

It is essential for intercomparing the results of different techniques that every technique uses the same system of constants. During the MERIT main campaign, all techniques adopted the MERIT Standards [Melbourne et al., 1983], so that the intercomparison may be carried out on the same theoretical basis. However, geopotential models, which are used by satellite techniques, are in different situations. For example, GEM-L2 model may be useful for the reduction of LAGEOS observations, but it is not so for TRANSIT satellite.

Establishment of the relativistic celestial frame is also an important task in consideration of the rapid improvement of observational accuracy [p.e. Fukushima et al., 1984].

The adopted constants are often improved by each technique to make them best fit to observations. This will reduce systematic errors. On the contrary, use of the stipulated constants will reduce systematic errors between different techniques, but the internal errors may increase. Refined models will also improve the celestial reference system.

Errors in the ERP's directly affect the estimation of the station coordinates. They give rise to errors in the direction of the baseline, but not in the baseline length in the case of a single determination. This is, however, not the case for a long series of data, because the errors of the ERP's cannot be constant throughout the long time span. Hence the baseline length is affected by the errors of the ERP's.

Importance of accurate determination of the perturbations of the Earth's rotation is not only in its geodynamical implications but also in its vital role for the precise determination of the contemporary plate motion.

References

Anderson, D. L., Earthquakes and the rotation of the earth, Science, 186, 49-50, 1974.

Anderson, D. L., and J. B. Minster, The frequency dependence of Q in the earth and implications for mantle rheology and Chandler wobble, Geophys. J. R. Astron. Soc., 58, 431-440, 1979.

Bender, P. L., and C. C. Goad, Probable LAGEOS contributions to a worldwide geodynamics control network, in The Use of Artificial Satellites for Geodesy and Geodynamics, 2, edited by G. Veis, and E. Livieratos, pp. 145-161, National Tech. Univ., Athens, 1979.

Bock, Y., The use of baseline measurements and geophysical models for the estimation of crustal deformations and the terrestrial reference system, Dept. Geod. Sci. Rept., 337, Ohio State Univ., 1982.

Carter, W. E., D. S. Robertson, J. E. Pettey, B. D. Tapley, B. E. Schutz, R. J. Eanes, and L. Miao, Variations in the rotation of the earth, Science, 224, 957-961, 1984.

Dahlen, F. A., The excitation of the Chandler wobble by earthquakes, Geophys. J. R. Astron. Soc., 25, 157-206, 1971.

Dahlen, F. A., A correction to the excitation of the Chandler wobble by earthquakes, Geophys. J. R. Astron. Soc., 32, 203-217, 1973.

Dahlen, F. A., The passive influence of the oceans upon the rotation of the earth, Geophys. J. R. Astron. Soc., 46, 363-406, 1976.

Elsasser, W. M., Sea-floor spreading as thermal convection, J. Geophys. Res., 76, 1101-1112, 1971.

Freeth, S. J., Can membrane tectonics be used to explain the break-up of plates?, in Mechanisms of Continental Drift and Plate Tectonics, edited by P. A. Davies and S. K. Runcorn, pp. 135-149, Academic Press, London, 1980.

Fricke, W., A rediscussion of Newcomb's determination of precession, Astron. Astrophys., 13, 298-308, 1971.

Fukushima, T., M. K. Fujimoto, K. Kinoshita, and S. Aoki, A system of astronomical constants in the relativistic framework, Cel. Mech., 38, 215-230, 1986.

Gilbert, F., and J. C. Dziewonski, An application of normal mode theory to the retrieval of structural parameters and source mechanisms from seismic spectra, Phil. Trans. R. Soc. London, 278, A.1280, 187-269, 1975.

Goldreich, P., and A. Toomre, Some remarks on polar wandering, J. Geophys. Res., 74, 2555-2567, 1969.

Guinot, B., A determination of the Love number k from the periodic waves of UT1, Astron. Astrophys., 36, 1-4, 1974.

Hales, A. L., Gravitational sliding and continental drift, Earth Planet. Sci. Lett., 6, 31-34, 1969.

Haubrich, R. A., An examination of the data relating pole motion to earthquakes, in Earthquake Displacement Fields and the Rotation of the Earth, edited by L. Manshinha, D. E. Smylie, and E. Beck, pp. 149-158, D. Reidel, Dordrecht Boston London, 1970.

Hess, H. H., History of ocean basins, in <u>Petrologic Studies: A Volume to Honour A. F. Buddington</u>, edited by A. E. J. Engel, H. L. James, and B. G. Leonard, pp. 599-620, Geod. Sci. Am., New York, 1962.

Hide, R., The earth's non-uniform rotation, in <u>Tidal Friction and the Earth's Rotation</u>, <u>2</u>, edited by P. Brosche, and J. Sündermann, pp. 92-97, Springer-Verlag, Berlin Heidlberg New York, 1982.

Hide, R., Excitation of short-term length-of-day changes and polar motion (abstract), <u>EOS Trans. AGU</u>, <u>65</u>, 186, 1984.

Hinderer, J., H. Legros, and M. Amalvict, A search for Chandler and nearly diurnal free wobbles using Liouville equations, <u>J. R. Astron. Soc.</u>, <u>71</u>, 303-332, 1982.

IRIS, <u>Earth Orientation Bulletin</u>, <u>3</u>, May, 1984.

Isacks, B., J. Oliver, and L. R. Sykes, Seismology and the new global tectonics, <u>J. Geophys. Res.</u>, <u>73</u>, 5855-5899, 1968.

Israel, M., A. Ben-Menahem, and S. J. Singh, Residual deformation of real earth models with application to the Chandler wobble, <u>Geophys. J. R. Astron. Soc.</u>, <u>32</u>, 219-247, 1973.

Jeffreys, H., The variation of latitude, <u>Mon. Not. R. Astron. Soc.</u>, <u>141</u>, 255-268, 1968.

Jeffreys, H., Creep in the earth and planets, in <u>Rotation of the Earth</u>, edited by P. Melchior, and S. Yumi, pp. 1-9, D. Reidel, Dordrecht Boston London, 1972.

Jurdy, D. M., and R. G. Gordon, Global plate motions relative to the hot spots 64 to 56 Ma, <u>J. Geophys. Res.</u>, <u>89</u>, 9927-9936, 1984.

Kimura, H., A new term in the variation of latitude: Independence of the components of the pole's motion, <u>Astron. J.</u>, <u>22</u>, 107-108, 1902.

Lambeck, K., <u>The Earth's Variable Rotation</u>, Cambridge University Press, 449 pp., 1980.

Langley, R. B., R. W. King, I. I. Shapiro, R. D. Rosen, and D. A. Salstein, Atmospheric angular momentum and the length of day: A common fluctuation with a period near 50 days, <u>Nature</u>, <u>294</u>, 730-732, 1981.

Le Pichon, X., Sea-floor spreading and continental drift, <u>J. Geophys. Res.</u>, <u>73</u>, 3661-3697, 1968.

Le Pichon, X., J. Francheteau, and J. Bonnin, <u>Plate Tectonics</u>, 300 pp., Elsevier, Amsterdam-London-New York, 1973.

Li, X. Q., and G. X. Song, Dynamo mechanism for turbulent wave in celestial bodies, <u>Astrophys. and Space Sci.</u>, <u>76</u>, 13-21, 1981.

Lieske, J. H., Precession matrix based on IAU (1976) System of Astronomical Constants, <u>Astron. Astrophys.</u>, <u>73</u>, 282-284, 1979.

Liu, H. S., On the breakup of tectonic plates by polar wandering, <u>J. Geophys. Res.</u>, <u>79</u>, 2568-2572 1974.

Lomnitz, C., Application of the logarithmic creep law to stress wave attenuation in the solid earth, <u>J. Geophys. Res.</u>, <u>67</u>, 365-368, 1962.

Luo, S. F., S. G. Liang, S. H. Ye, S. Z. Yan, and Y. X. Li, Analysis of the periodicity of the irregular rotation of the earth (in Chinese), <u>Acta Astron. Sinica</u>, <u>15</u>, 79-85, 1974.

Manshina, L., and D. E. Smylie, Effect of earthquakes on the Chandler wobble and the secular polar shift, <u>J. Geophys. Res.</u>, <u>72</u>, 4731-4743, 1967.

Mansinha, L., D. E. Smylie, and C. H. Chapman, Seismic excitation of the Chandler wobble revisited, <u>Geophys. J. R. Astron. Soc.</u>, <u>59</u>, 1-17, 1979.

Markowitz, W., Sudden changes in rotational acceleration of the earth and secular motion of the pole, in <u>Earthquake Displacement Fields and the Rotation of the Earth</u>, edited by L. Manshinha, D. E. Smylie, and A. E. Beck, pp. 69-81, D. Reidel, Dordrecht Boston New York, 1970.

Melbourne, W., R. Anderle, M. Feissel, R. King, D. McCarthy, D. Smith, B. Tapley, and R. Vicente, Project MERIT standards, <u>USNO Circular</u>, <u>167</u>, 1983.

Minster, J. B., and T. H. Jordan, Present-day plate motions, <u>J. Geophys. Res.</u>, <u>83</u>, 5331-5354, 1978.

Morgan, W. J., Rises, trenches, great faults, and crustal blocks, <u>J. Geophys. Res.</u>, <u>73</u>, 1959-1982, 1968.

Morgan, W. J., Deep mantle convection plumes and plate motions, <u>Am. Assoc. Petrol. Geol. Bull.</u>, <u>56</u>, 203-213, 1972.

Moritz, H., Theories of nutation and polar motion 1, <u>Dept. Geod. Sci. Rept.</u>, <u>309</u>, Ohio State Univ., 1980.

Mueller, I. I., Reference coordinate system for earth dynamics: A preview, in <u>Reference Coordinate Systems for Earth Dynamics</u>, edited by E. M. Gaposchkin, and B. Kołaczek, pp. 1-22, D. Reidel, Dordrecht Boston London, 1981.

Mueller, I. I., S. Y. Zhu, and Y. Bock, Reference frame requirements and the MERIT campaign, <u>Dept. Geod. Sci. Rept.</u>, <u>329</u>, Ohio State Univ., 1982.

Munk, W. H., and G. J. F. MacDonald, <u>The Rotation of the Earth</u>, 323 pp., Cambridge University Press, 1960.

Musman, S., Statistical tests of ARIES data, <u>J. Geophys. Res.</u>, <u>87</u>, 5553-5562, 1982.

Myerson, R. J., Long-term evidence for the Association of earthquakes with the excitation of the Chandler wobble, <u>J. Geophys. Res.</u>, <u>75</u>, 6612-6617, 1970.

O'Connell, R. J., and A. M. Dziewonski, Excitation of the Chandler wobble by large earthquakes, <u>Nature</u>, 259-262, 1976.

Okamoto, I., and N. Kikuchi, Low frequency variations of homogeneous ILS polar motion data, <u>Publ. Intern. Latitude Obs. Mizusawa</u>, <u>16</u>, 35-40, 1983.

Okubo, S., Theoretical and observed Q of the Chandler wobble - Love number approach, <u>Geophys. J. R. Astron. Soc.</u>, <u>71</u>, 647-657, 1982.

Poma, A., and E. Proverbio, Astronomical evidence of relationships between polar motion, earth rotation and continental drift, in <u>Mechanisms</u>

of Continental Drift and Plate Tectonics, edited by P. A. Davies, and S. K. Runcorn, pp. 345-357, Academic Press, London, 1980.

Proverbio, E., and V. Quesada, Secular variations in latitudes and longitudes and continental drift, J. Geophys. Res., 79, 4941-4943, 1974.

Richter, F. M., Dynamical models for sea floor spreading, Rev. Geophys. Space Phys., 11, 223-287, 1973.

Richter, B., Concepts of reference frames for a deformable earth, in Reference Coordinate Systems for Earth Dynamics, edited by E. M. Gaposchkin, and B. Kołaczek, pp. 261-265, D. Reidel, Dordrecht Boston London, 1981.

Rochester, M., The effect of the core on the earth's rotation, Veröff. Zentralinst. Phys. Erde, 30, 77-89, Akad. Wiss. DDR, 1974.

Rosen R. D., and D. A. Salstein, Variations in atmospheric angular momentum, Environmental Research & Technology, Inc. Rept., A345-T1, Concord, MA, 1981.

Rosen, R. D., D. A. Salstein, T. M. Eubanks, J. O. Dickey, and J. A. Steppe, An El Niño signal in atmospheric angular momentum and earth rotation, Science, 225, 411-414, 1984.

Sasao, T., I. Okamoto, and S. Sakai, Dissipative core-mantle coupling and nutational motion of the earth, Publ. Astron. Soc. Japan, 29, 83-105, 1977.

Sasao, T., S. Okubo, and M. Saito, Dynamical effects of stratified fluid core upon nutational motion of the earth, in Nutation and the Earth's Rotation, edited by E. P. Fedorov, M. L. Smith, and P. L. Bender, pp. 165-183, D. Reidel, Dordrecht Boston London, 1979.

Sato, T., Nearly diurnal free core resonance measured by extensometers, in Proc. Japanese Symp. on Application of Space Techniques for Geodesy and Geodynamics, edited by S. Aoki, pp. 547-555, Tokyo University Press, 1984.

Slade M. A., G. A. Lyzenga, and A. Raefsky, Interaction of mantle rheology and plate boundary earthquakes in excitation of Chandler wobble (abstract), EOS Trans. AGU, 65, 187, 1984.

Smith, M. L., Wobble and nutation of the Earth, Geophys. J. R. Astron. Soc., 50, 103-140, 1977.

Smith, M. L., The theoretical description of the nutation of the earth, in Reference Coordinate Systems for Earth Dynamics, edited by E. M. Gaposchkin, and B. Kołaczek, pp. 103-110, D. Reidel, Dordrecht Boston London, 1981.

Smith, M. L., and F. A. Dahlen, The period and Q of the Chandler wobble, Geophys. J. R. Astron. Soc., 64, 223-281, 1981.

Smylie, D. E., and L. Mansinha, Earthquakes and the observed motion of the rotation pole, J. Geophys. Res., 73, 7661-7673, 1968.

Smylie, D. E., and L. Manshinha, The elasticity theory of dislocations in real earth models and changes in the rotation of the earth, Geophys. J. R. Astron. Soc., 23, 329-354, 1971.

Song, G. X., On the influence of polar shift upon tectonic plate boundaries (in Chinese), Ann. Shanghai Observatory, 1, 15-19, 1979.

Song, G. X., M. Zhao, and D. W. Zheng, On the excitation of earthquake displacement field to polar motion (in Chinese), Acta Astron. Sinica, 22, 383-388, 1981.

Spencer-Jones, H., The rotation of the earth, and the secular accelerations of the sun, moon and planets, Mon. Not. R. Astron. Soc., 99, 541-558, 1939.

Stephanson. F. R., J. V. Morrison, and G. A. Wilkins, Secular and decade changes in the lod, in Proc. IAG Symp., 2, pp. 62-64, XVIII IUGG Gen. Assem., Hamburg, August 15-27 1983, Dept. Geod. Sci. Surv., Ohio State Univ., 1983.

Stefanick, M., and M. Jurdy, The distribution of hot spots, J. Geophys. Res., 89, 9919-9925, 1984.

Turcotte, D. L., and E. R. Oxburgh, Mid-plate tectonics, Nature, 244, 337-339, 1973.

Wahr, J., A normal mode expansion for the forced response of a rotating earth, Geophys. J. R. Astron. Soc., 64, 651-675, 1981.

Walter, L. S., Geodynamics, NASA Conference Publication, 2325, 49 pp., 1984.

Wells, J. W., Paleontological evidence of the rate of the earth's rotation, in The Earth-Moon System, edited by B. G. Marsden, and A. G. W. Cameron, pp. 70-81, Plenum Press, New York, 1966.

Wilkins, G. A., Report of the 2nd MERIT Workshop, Royal Greenwich Observatory, 1984.

Woolard, E. W., Theory of the rotation of the earth around its center of mass, Astron. Papers, 15, Part 1, 1-165, Nautical Almanac Office, USNO, 1953.

Yumi, S., and K. Yokoyama, Results of the ILS in a homogeneous system: 1899.9-1979.0, 198 pp., Central Bureau of the IPMS, Mizusawa, 1980.

Zhao, M., and D. W. Zheng, On the discussion of secular polar motion (in Chinese), Acta Astron. Sinica, 21, 69-72, 1980

Zheng, D. W., An analysis of the short period terms in the universal time (in Chinese), Acta Astron. Sinica, 19, 103-108, 1978.

Zheng, D. W., and M. Zhao, Application of Autoregressive technique to astronomy and geodynamics (in Chinese), Acta Astron. Sinica, 20, 301-307, 1979.

Zhu, S. Y., Some problems in setting up the new conventional terrestrial system (in Chinese), Acta Astron. Sinica, 24, 327-331, 1983.

Zhu, S. Y., and I. I. Mueller, Effects of adopting new precession, nutation and equinox corrections on the terrestrial reference frame, Bull. Geod., 57, 29-42, 1983.

RECENT DEVELOPMENTS IN SPACE GEODESY: EDITORIAL NOTE

Keichi Kasahara

Earthquake Research Institute, University of Tokyo, Japan[1]

Rapid progress in space geodesy has continued, even during the course of editing this volume. Three recent developments should be mentioned:

Motion of the Pacific Plate

The US-Japan joint VLBI experiments (NASA and the Radio Research Laboratory) have repeated measurement of baseline lengths in and around the Pacific Ocean, several times since 1984. Figure 1 (Kondo et al., 1986) illustrates the motion of the Pacific stations relative to Kashima, Japan, as observed in 1984-85. The motions of the other stations in the same period are also given for reference.

The formal error of distance measurements was less than 3 cm in every experiment, which allows us to study the relative motion of the stations with sufficiently high precision. The measured distance between Kashima and Kauai, for example, decreased by 6.9 ± 1.9 cm/year, whereas a long-term convergence rate of 7.7 cm/year is theoretically expected. The agreement of theory and experiment is generally good for every baseline, as seen in the figure. This is a preliminary result. With this type of data, we can reach our primary goal of direct measurement of on-going plate motion.

Regional and Local Space Geodesy

Progress in space techniques has opened a new era of geodesy involving intercontinental networks of SLR and VLBI stations. Another notable feature of this new era is developing technology for regional and local space geodesy, involving mobilized SLR and VLBI systems. Among the new concepts proposed are, satellite borne laser or Doppler systems, and precise orbiting of satellites. Geodetic applications of Global Positioning System (GPS) are also of wide interest.

The GPS method is basically a phase comparison of signals from space at a pair of stations. Use of a small omni-directional antenna, instead of the large parabola antenna required for the VLBI, allows this system to be highly mobile and relatively inexpensive. It is extremely advantageous for geologists and geophysicists studying local and regional crustal strain accumulation in tectonically active areas. Tens of groups, both academic and industrial, are now testing various models of this system in the US, Europe, Japan and other regions (e.g. Davidson et al (1986), also, CSTG (1985)).

The International Earth Rotation Service

A new plan for the international Earth Rotation Service (IERS) has been proposed by an international group of astrogeodesists (Wilkins and Mueller, 1986). It will coordinate international observations of the earth-rotation parameters using modern techniques. The principal features of the proposed organization, which are explained in the report by Wilkins and Mueller, are as follows:

(1) A Central Bureau that would receive rapid and final results from the coordinating and analysis centres and would derive and disseminate earth-rotation parameters and information relating to the reference frames.

(2) Coordinating Centres, each of which would be responsible for coordinating the observational and data processing activities associated with one technique or task. Within a given technique it may be necessary to include operational network centres and analysis centres in order to spread the load of work and make the best use of the available expertise.

(3) "Observatories" that would be willing to supply observational data regularly to the appropriate coordinating centre. (An "observatory" in this context could be a network of stations administered by a single organization.)

(4) A Directing Board that would consist of representatives of the Unions and of the Central Bureau and Coordinating Centres and that would exercise general administrative and technical supervision of the activities of the Service, and would ensure that it continues to meet the changing needs of the wide user community.

---

[1] Present address: Science & Engineering Research Laboratory, Waseda University, Kikui-cho, Shinjuku-ku, Tokyo 162, Japan

Copyright 1987 by the American Geophysical Union.

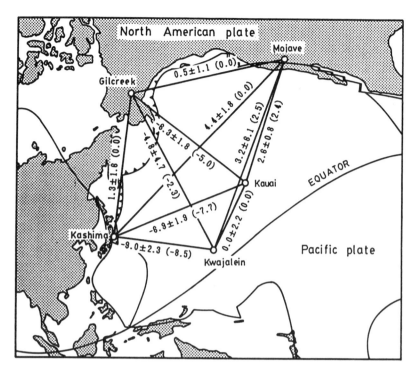

Fig. 1. Observed and expected rates of baseline length change (cm/year). Numerals in brackets are the rates as theoretically expected. After Kondo et al. (1986).

This Service is expected to start in 1988, after approval by the IAU and IUGG.

References

CSTG (International Coordination of Space Techniques for Geodesy and Geodynamics), Future Missions, Systems and Projects, CSTG Bulletin, No. 8, pp. 230, 1985.

Davidson, J.M., T.H. Dixon, M.P. Golombek, W.G. Melbourne, C.L. Thornton, and C.J. Vegos, GPS Measurement for Regional Geodesy, Operating Plan, JPL Report 1710-5, pp. 49, 1986.

Kondo, T., K. Heki, and Y. Takahashi, Pacific Plate Motion Detected by the VLBI Experiments Conducted in 1984-85, J. Radio Res. Lab., (in press), 1986.

Wilkins, G.A., and I.I. Mueller, On the Rotation of the Earth and the Terrestrial Reference System: Joint Summary Report of the IAU/IUGG Working Groups MERIT and COTES, Bull. Géod., 60, 85-100, 1986.

TRENCH DEPTH AND RELATIVE MOTION BETWEEN OVERRIDING PLATES

Kazuaki Nakamura

Earthquake Research Institute, University of Tokyo, Tokyo 113

Abstract. According to recent research, trench depth varies with the age of the subducting plate and with the rate of convergence. It is shown here that the deepening from the Japan trench to the Izu·Bonin trench at the TTT triple junction can not be explained by the above two factors, but is explained by the retreating motion of the Philippine Sea plate, one of the two overriding plates, relative to the other, the northeast Honshu plate. The depth change of the Middle America trench at the Cocos-North America-Caribbean TTF triple junction might be another example of the same mechanism. Possibilities for explaining the existence of unstable triple junctions, which are discussed briefly, include: tectonic erosion of the non-retreating overriding plate, rifting and thinning of the retreating overriding plate, and lateral accretion of ultramafic rocks supplied from the mantle wedge along the trench side margin of the retreating plate.

Introduction

The idea that oceanic trenches are the topographic expression of the down bending of oceanic plates associated with subduction is well established. However, the significance of variation in trench depth has only been addressed recently by Hilde and Uyeda (1981, 1983) and by Grellet and Dubois (1982). According to these authors, trench depth is greater with older ages of the subducting slab (Hilde and Uyeda, 1983) and with higher rate of convergence (Grellet and Dubois, 1982). This is consistent with the general correlation between the age and velocity of subducting plate (e.g. Carlson et al., 1983; Peterson and Seno, 1984; Hilde and Uyeda, 1983).

In this article, it is shown that at least in some cases the relative motion between the overriding plates affects the trench depth apparently more decisively than both the age of the subducting slab and the convergence rate. In this context, the reasons for the existence of theoretically unstable triple junctions are briefly sought.

Copyright 1987 by the American Geophysical Union.

Japan-Izu·Bonin Trenches

The Pacific (PAC) plate subducts westwards along the Japan trench beneath Northeast Honshu and along the Izu·Bonin trench beneath the Philippine Sea (PHS) plate (Figure 1). The two trenches are separated geographically (Hydrogr. Dept., 1982) by a seamount, Kashima No.1 seamount, that is located on the trench axis, and is broken and about to be subducted (Mogi and Nishizawa, 1980). In this article, the boundary between the Japan and Izu·Bonin trenches is taken, for convenience, as the triple junction ca 180 km south of the Kashima No.1 seamount.

The variation in depth along the axis of the trench is shown in Figure 2. The Japan trench is shallower, about 7.5 km deep, and the Izu·Bonin trench is depper, about 9 km or more deep. There is an offset of about 1.5 km between the depth of the two trenches at the trench triple junction where the Sagami trough joins the other two trenches. The Sagami trough is a shallow trench and is a highly oblique subduction boundary where the PHS plate subducts beneath NE Honshu. Therefore, the trench junction is actually a trench triple junction. The abrupt change in depth between the Japan and the Izu·Bonin trenches is accompanied by a change in the width of the trenches as defined by the distance between 6,000 m depth contour lines. This is also shown as thinly stippled areas in Figure 1 where the widths of the Japan and the Izu·Bonin trenches are 70±10 km and 90±10 Km, respectively.

The observation that the Izu·Bonin trench is deeper than the Japan trench is reinforced by the observation that the shallower Japan trench is almost starved of trench-fill sediments whereas the deeper Izu·Bonin trench is widely, therefore thickly, filled with sediments (Geol. Surv. Japan, 1982; Renard, Nakamura et al., 1987). These trench-fill sediments are now known to have been supplied through Boso canyon in the Sagami trough from colliding and uplifting volcanic central Japan (Kato et al., 1985; Renard, Nakamura et al., 1987). Both trenches deepen slightly towards south. Southward deepening of the Japan trench, which is not obvious in Figure 2, was recently confirmed

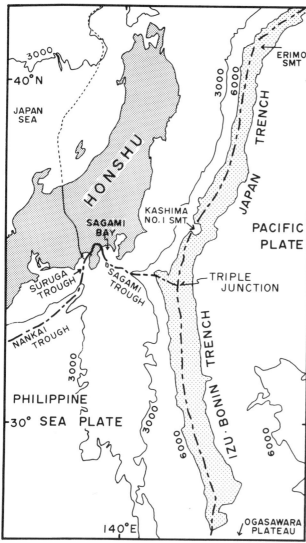

Fig. 1. Trench-trench-trench triple junction off Central Japan. Difference in the width between the Japan and the Izu·Bonin trenches is shown by stippled area deeper than 6 km.

by Cadet, Kobayashi et al. (1987). According to them, the trench axis deepens southwards by 0.6 km between 40°N and 36°N. The average depth is 7.5 km.

The observation that the turbidites filling the Izu·Bonin trench are supplied at the triple junction explains, at least partly, why the Izu·Bonin trench deepens southward away from the triple junction. Similar unidirectional deepening is observed along the Nankai trough and the Sunda trench which are filled with sediments supplied from central Japan and eastern Himalayan mountains, respectively (Taira and Niitsuma, 1986; Moore et al., 1982).

The age of the subducting Pacific plate changes gradually, not abruptly along the two trenches: 125±5 Mabp for the Japan trench and 135±5 Mabp for the Izu·Bonin trench (Hilde et al., 1976). The possible 10 Ma difference in the average age of the subducting plate is, however, not enough to account for the 1.5 km depth difference between the two trenches as the age effect (Hilde and Uyeda, 1983). The slight southward deepening tendency as observed for both trenches may be related partly to the southward increasing age of the subducting PAC plate along both trenches.

The greater depth of the Izu·Bonin trench relative to the Japan trench is just opposite to what is expected from the comparison of the convergence rate, which is faster by 3 cm/a along the shallower Japan trench than along the deeper Izu·Bonin trench (Minster and Jordan, 1978, 1979). This rate difference is qualitatively obvious also from the observation that the PHS plate has no significant spreading center within it and is subducting along its western margin beneath Japanese islands (Circum Pacific Council for Energy and Mineral Resources, 1981a). Thus abrupt depth variation across the triple junction is not explained in terms of the two factors already established: the age of the subducting plate and the convergence rate.

Hypothesis

It is proposed here that the adrupt change in the trench depth is due to the differential movement between two overriding plate.

The systematic difference in the depth of Japan and Izu·Bonin trenches coincides well geographically with the abrupt change at the triple junction in the motion between the two overriding plates. The PHS plate is subducting beneath northeast Honshu obliquely in northwest direction along the Sagami trough (Figure 3) as evidenced by the 1923 Kanto earthquake (M 8.2, Kanamori, 1971; Ishibashi, 1981) as well as geological and geomorphological observations in the Sagami trough (Nakamura et al., 1984, Nakamura, Renard et al., 1987). Also, the relative motion models of rigid plate tectonics (e.g. Ranken et al., 1984; Seno, 1977) show that the PHS plate is moving northwest

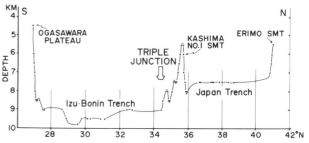

Fig. 2. Longitudinal variation in the axial depth of the Japan and the Izu Bonin trenches.

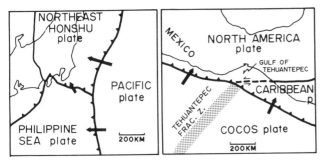

Fig. 3. Plate tectonic setting of TTT triple junction off Central Japan and TTF triple junction in the Middle America trench. Arrows: approximate convergence vector (CPC EMR, 1981a; Burback et al., 1984).

relative to northeast Honshu with a component of westward motion of a few cm per year.

On the other hand, the PAC plate is subducting westwards along the Japan-Izu·Bonin trenches beneath northeast Honshu in the north and beneath PHS plate in the south. The subducted PAC slab is laterally continuous all along the trenches as evidenced by a continuous Wadati-Benioff seismic plane without major segmentation, although the general dip of the slab is much steeper in the south (Utsu, 1984, p.158; Yoshii, 1979). Therefore, the ocean side wall of the Izu·Bonin trench should have moved together with that of the Japan trench. This means that the location of the Izu·Bonin trench should have been more or less fixed with respect to the Japan trench, regardless of the motion of the PHS plate. Then, the existence of the component of the westward motion of the PHS plate relative to the northeast Honshu will necessarily result in the retreat of the overriding PHS plate away from the Izu·Bonin trench causing its deepening and widening.

It should be noted that a similar idea to explain the difference in trench depth north and south of the TTT junction was expressed in Renard, Nakamura et al. (1987).

Another Possible Example-Middle America Trench

In the central part of the 3,000 km long Middle America trench there is a plate tectonic setting which is almost a mirror image of the triple junction off central Japan (Figure 3). The Cocos plate subducts northeastward all along the Middle America trench (Molnar and Sykes, 1969). The overriding plates are the North America plate to the northwest and the Caribbean plate to the southeast. The left-lateral strike-slip boundary between the two plates, the Polochic-Motagua fault zone, joins the trench to form a TTF triple junction off the Gulf of Tehuantepec. Because the Caribbean plate is moving east relative to the North America plate (e.g. Sykes et al., 1982), the TTF triple junction in the Middle America trench is similar in plate tectonic setting to the TTT junction off Japan, in that one of the overriding plate is retreating with respect to the other overriding plate.

The depth and width of the Middle America trench varies just as is expected from the hypothesis stated above. The trench deepens abruptly toward southeast within 60 km by 500 fathoms (about 0.9 km), from 2,800 fathoms to 3,300 fathoms, at the triple junction (Fisher, 1961). The variation from a shallower northwestern part to a deeper southeastern part is conspicuous in the change in width of the trench. When we take the distance between the 2,500 fathom contour lines as representing the trench width, it is about 20 km to the northwest and about 40 km to the southeast. The widening takes place within some 30 km along the trench in the area of the triple junction (Fisher, 1961). The rate of convergence along the Middle America trench increases toward southeast approximately from 6 to 8 cm/a (Minster and Jordan, 1978). At the TTF triple junction, however, it decreases from 7.8 cm/a to 6.9 cm/a (Burbach et al., 1984). The convergence rate is lower for the deeper trench and this is opposite from what is expected from the established rate-depth relation. As with the PAC plate, the subducted Cocos plate is without any major discontinuity, and has a steeper dip to the southeast of the triple junction (Burbach et al., 1984).

Thus, it appears plausible that the same mechanism that governs the trench depth off Central Japan applies to the trench-depth variation in the Middle America trench, although there remains a possibility that the depth variation may mostly be explained by the age and the depth of the subducting Cocos plate. The Tehuantepec fracture zone is subducting in the area of the triple junction (Burbach et al., 1984), and the depth of the ocean floor before subduction is shallower by about 0.6 km to the northwest of the fracture zone than to the southeast (Fisher, 1961; Mammerickx and Smith, 1982). Thus the deeper ocean floor may contribute at least partially to the abrupt deepening of the southeastern part of the trench. Another point is that the age of the subducting plates is probably more than several Ma older (Truchan and Larson, 1973; Klitgord and Mammerickx, 1982) for the southeastern deeper part than the northwestern part. Thus, the older age of the southeastern part correlates with the greater depth of the trench to the southeast, suggesting the possibility that the greater depth may at least be partially accounted for by the older age.

Discussion

Because the two triple junctions with which we are concerned are unstable under the assumption of rigid plate tectonics (McKenzie and Morgan, 1969), their present existence requires some explanation. There are at least five possibilities: 1) the relative motion between the two overriding plates has just started; 2) the retreating part of the trench has just arrived at the present position; 3) the edge of the non retreating, overriding plates has been continuously eroded tectonically; 4) a substantial amount of the expected retreat has been accommodated by thinning of the retreating plate; and 5) along the retreating margin of the overriding plate ultramafic material is being accreted laterally to keep the location of the trench fixed. Possibilities 1, 2 and 5 fit within the geometrical framework of rigid plate tectonics, while possibilities 3 and 4 do not.

Possibility 1: that the relative motion between the two overriding plates started so recently as to make the cumulative amount of displacement insignificant, can not be accepted. A long history of relative motion between the North American and Caribbean plates is well documented (e.g. Sykes et al., 1982). The present phase of the northwestward motion of the PHS plate relative to northeast Honshu island is thought to have started during middle Quaternary (Nakamura et al., 1984; Kaizuka, 1984) or earlier according to studies of the age, sense and strike of faults and other geologic structures of Honshu. Therefore, at least a few tens of kilometers of misfit of the trench axis is expected. Whereas, detailed observations of the trench (Renard, Nakamura et al., 1987) do not fit this expectation.

Possibility 2: that the present configuration has just formed because the trench beside the retreating plate just arrived on the extension of the trench beside the non retreating plate to form a single linear trench, may not be dismissed completely, but is not appealing because it depends too much on chance. Moreover, the lateral continuity of the subducted PAC and Cocos plates makes the interpretation even more implausible.

The third possibility is that the present configuration of the triple junctions is the result of tectonic erosion of the edge of the non-retreating plate. Carlson and Melia (1984) postulated the advance of the hinge of the subducting PAC plate, or roll foreward, along the Izu·Bonin trench on the basis of the absence of active backarc spreading within the retreating PHS plate. In the context of the present discussion, the advance of the Izu·Bonin trench is almost equivalent to the tectonic erosion of the overriding, non-retreating plate, along the Japan trench beside northeast Honshu. This mechanism was first proposed for the Japan trench off Central Japan by Murauchi and Asanuma (1970) and developed by Sugimura (1972) and Matsubara and Seno (1980). Their primary observation is that the geological structures of the edge of the overriding plate seems to be cut morphologically by the Japan trench (Murauchi and Asanuma, 1970). However, this explanation also appears unfavorable from observation of the current geographical configuration of the trench, because it requires that the edge of the non-retreating plate was removed so that a sharp bend in arc shape would have been resulted for that part of the trench.

The present configuration of the Japan trench and the northwestern part of the Middle America trench are both essentially linear (Figure 3). Although this explanation does not appear favorable, the possibility that tectonic erosion occurred in the edge of the non-retreating plate cannot be ruled out.

The fourth possibility is that the retreating motion has been accommodated by the thinning of the retreating overriding plate. This sort of phenomena was documented by Mann and Burke (1984) and Plafker (1976) for the northwestern part of the Caribbean plate as several north-south striking rift structures. For the PHS plate, Tamaki et al., (1981) and Honza and Tamaki (1985) found young back arc rift immediately west of the Izu volcanic line along 600 km between 27°N and 33°N, and Yuasa (1985) documented north-south striking normal faults cutting the sea floor in the forearc region. This rifting and normal faulting has certainly accommodated part of the retreating motion (Carlson and Melia, 1984); but it is not certain whether or not it has accommodated all of it.

The fifth possibility is that along the zone immediately landward of the trench, retreating motion of the overriding plate has been accommodated by intrusion of mafic and ultramafic rocks supplied from the wedge mantle of the overriding plate. This possibility is suggested by the occurrence in the trench inner wall of ultramafic rocks of island arc origin for both Izu·Bonin trench (Ishii, 1985: Taylor and Smoot, 1984; Taylor personal communication, 1985) and the southeastern part of the Middle America trench (Aubouin et al., 1982) and by the non occurrence of such ultramafic rocks along both the Japan trench (Honza, 1977) and the northwestern part of the Middle America trench (Moore et al., 1979). In the case of the Middle America trench, however, the age of the "ophiolite" (Aubouin et al., 1982) is judged as pre-Eocene based on the age of the overlying sediments. It may still be possible that these ultramafic rocks have been somehow underplated after the deposition of the overlying sediments without much disturbing them. Although the mechanism to supply enough area to compensate the retreat of the plate remains unknown, this possibility may be tested by geologic mapping of the forearc region close to the trench to see if a correlation exists between the retreat and the occurrence of ultramafic intrusions for other arcs

Concluding Remarks

Nothing can be said definitely about the reason why we can observe at present the alignment of the retreating and non-retreating trench axis segments at the two triple junctions discussed above. Possibilities 3, 4 and 5: tectonic erosion, thinning of the retreating plate and possible accretion of ultramafic rocks along the trench, deserve further research.

Nevertheless, the sense of relative motion between two overriding plates seems to be affecting the depth of the trench on both sides of the TTT triple junction more decisively than both the age of the subducting plate and convergence rate at least in some cases discussed in this paper.

Acknowledgement. The earlier manuscript was critically and helpfully reviewed by Seiya Uyeda and Raymond Price.

References

Aubouin, J., R. von Huene, M. Baltuck, R. Arnott, J. Bourgois, M. Filewicz, K. Kvenvolden, B. Leinert, T. McDonald, K. McDougal, Y. Ogawa, E. Taylor, and B. Winsborough, Leg 84 of the Deep Sea Drilling project, Nature, 297, 458-460, 1982.

Burbach, G. V., C. Frohlich, W. D. Pennington, and T. Matumoto, Seismicity and tectonics of the subducted Cocos plate, J. Geophys. Res., 89, 7719-7735, 1984.

Cadet, J. P., K. Kobayashi, J. Aubouin, J. Boulegur, J. Dubois, R. von Huene, L. Jolivet, T. Kanazawa, J. Kasahara, K. Koizumi, S. Lallemand, Y. Nakamura, G. Pautot, K. Suyehiro, S. Tani, H. Tokuyama, and T. Yamazaki, The Japan trench and its juncture with the Kuril trench. Cruise results of the KAIKO project, Leg. 3, Earth Planet. Sci. Lett., 1986, in press.

Carlson, R. L., and P. J. Melia, Subduction hinge migration, Tectonophys., 102, 399-411, 1984.

Carlson, R. L., T. W. C. Hilde, and S. Uyeda, The driving mechanism of plate tectonics: Relation to age of the lithosphere at trenches, Geophys. Res. Lett., 10, 297-300, 1983.

Circum-Pacific Council for Energy and Mineral Resources, Plate tectonic map of the Circum Pacific Region, northwest quadrant, AAPG, 1981a.

Circum-Pacific Council for Energy and Mineral Resources, Plate tectonic map of the Circum Pacific Region, northeast quadrant, AAPG, 1981b.

Fisher, R. L., Middle America trench: Topography and structure, Geol. Soc. Amer. Bull., 72, 703-720, 1961.

Geological Survey of Japan, Geological map of the northern Ogasawara arc, Marine Geology Map Series, 17, 1982.

Grellet, C., and J. Dubois, The depth of trenches as a function of the subduction rate and age of the lithosphere, Tectonophys., 82, 45-56, 1982.

Hilde, T. W. C., N. Isezaki, and J. M. Wageman, Mesozoic Sea-floor spreading in the North Pacific, Geophys. Monogr. Ser. A.G.U., 19, 205-226, 1976.

Hilde, T. W. C., and S. Uyeda, Trench depths vs. age of subducting plate, read before OJI seminar on accretion tectonics, Tomakomai, Japan, 1981.

Hilde, T. W. C., and S. Uyeda, Trench depth: Variation and significance, Geodynamic Ser., Amer. Geophys. Union, 11, 75-89, 1983.

Honza, E. (ed.), Geological investigation of Japan and southern Kuril trench and slope areas, Cruise Rept. Geol. Surv. Japan, n.7, 1-127, 1977.

Honza, E., and K. Tamaki, The Bonin arc, in Ocean basins and margins, Plenum. V7A, edited by A. E. M. Nairn, and S. Uyeda, pp.459-502, 1985.

Hydrographic Department, M.S.A., Bathymetric chart, n.6313, Central Nippon 1/1,000,000, 1982.

Ishibashi, K., Fault plane solution of 1923 Kanto earthquake (in Japanese), Chikyu, 3, 452-454, 1981.

Ishii, T., Dredged samples from the Ogasawara forearc seamount or Ogasawara Paleoland - forearc ophiolite, in Formation of Active Margins, edited N. Nasu, and I. Kushiro, Terra Sci. Publ. Co., Tokyo, 1985

Kanamori, H., Faulting of great Kanto earthquake of 1923 as revealed by seismological data, Bull. Earthq. Res. Inst. Univ. Tokyo, 49, 13-18, 1971.

Kato, S., T. Nagai, et al., Submarine topography of the eastern Sagami trough to the triple junctions, Rpt. Hydrogr. Res., 20, 1-24, 1985.

Klitgord, K. D., and J. Mammerickx, Northern East Pacific rise: Magnetic anomaly and bathymetric framework, J. Geophys. Res., 87, 6725-6750, 1982.

Mammerickx, J., and S. M. Smith, General Bathymetric Chart of the Oceans, 1/1,000,000, 5 07, Canadian Hydrographic Service, Ottawa, Canada, 1982.

Mann, P., and K. Burke, Cenozoic rift formation in the northern Caribbean, Geology, 12, 732-736, 1984.

Matsubara, Y., and T. Seno, Paleongeographic reconstruction of the Philippine Sea at 5 m.y. B.P., Earth Planet. Sci. Lett., 51, 406-414, 1980.

McKenzie, D. P., and W. J. Morgan, Evolution of triple junctions, Nature, 224, 125-133, 1969.

Minster, J. B., and T. H. Jordan, Present-day plate motions, J. Geophys. Res., 83, 5331-5354, 1978.

Minster, J. B., and T. H. Jordan, Rotation vectors for the Philippine and Rivera plates, EOS, 60, 958, 1979.

Mogi, A., and K. Nishizawa, Breakdown of a seamount on the slope of the Japan trench, Proc. Japan Acad., 56, Ser. B., 257-260, 1980.

Molnar, P., and L. Sykes, Tectonics of the Caribbean and Middle America regions from focal mechanisms and seismicity, Geol. Soc. Am. Bull., 80, 1639-1684, 1969.

Moore, G. F., J. R. Curray, and F. J. Emmel, Sedimentation in the Sunda trench and forearc region, in Trench-foearc geology: Sedimentation and tectonics on modern and ancient active plate margins, edited by Leggett, pp.245-258, 1982.

Moore, J. C., J. C. Watkins, T. H. Shipley, S. B. Backman, F. W. Bcthtel, A. Butt, B. M. Didyk,

Leggett, J. K., N. Lundberg, K. J. McMillen, N. Niitsuma, L. E. Shepard, F. J. Stephan, and H. Stradner, Progressive accretion in the Middle America trench, southern Mexico, Nature, 281, 638-642, 1979.

Murauchi, S., and T. Asanuma, Studies on seismic profiler measurements off Boso-Jyoban district, northeast Japan, Bull. Nat. Sci. Mus. Tokyo, 13, 337-355, 1970.

Nakamura, K., V. Renard, J. Angelier, J. Azema, J. Bourgois, C. Deplus, K. Fujioka, Y. Hamano, P. Huchon, H. Kinoshita, P. Labaume, Y. Ogawa, T. Seno, A. Takeuchi, M. Tanahashi, A. Uchiyama, and J.-L. Vigneresse, Oblique and near collision subduction, Sagami and Suruga troughs - Preliminary results of French-Japanese 1984 KAIKO cruise, leg. 2., Earth Planet. Sci. Lett., 1987, in press.

Nakamura, K., K. Shimazaki, N. Yonekura, Subduction, bending and eduction, Present and Quaternary tectonics of the northern border of the Philippine sea plate, Bull. Soc. Geol. France, 26, 221-243, 1984.

Peterson, E. T., and T. Seno, Factors affecting seismic moment release rates in subduction zones, J. Geophys. Res., 89, 10233-10248, 1984.

Plafker, G., Tectonic aspects of the Guatemala earthquake of 4 February 1976, Science, 193, 1201-1208, 1976.

Ranken, B., R. K. Cardwell, and D. E. Karig, Kinematics of the Philippine Sea plate, Tectonics, 3, 555-575, 1984.

Renard, V., K. Nakamura, J. Angelier, J. Azema, J. Bourgois, C. Deplus, K. Fujioka, Y. Hamano, P. Huchon, H. Kinoshita, P. Labaume, Y. Ogawa, T. Seno, A. Takeuchi, M. Tanahashi, A. Uchiyama, and J.-L. Vigneresse, Trench triple junction off central Japan - Preliminary results of French-Japanese 1984 KAIKO cruise, leg. 2., Earth Planet. Sci. Lett., 1987, in press.

Seno, T., The instantaneous rotation vector of the Philippine Sea plate relative to the Eurasian plate, Tectonophys., 42, 209-226, 1977.

Sugimura, A., Plate boundaries in and around Japan, Kagaku, 42, 192-202, 1972.

Sykes, L. R., W. R. McCann, and A. L. Kafka, Motion of Caribbean plate during last 7 million years and implications for earlier Cenozoic movements, J. Geophys. Res., 87, 10656-10676, 1982.

Taira, A., and N. Niitsuma, Turbidite sedimentation in Nankai trough as interpreted from magnetic fabrics, grain size and detrital modal analyses, DSDP-IPOD leg 87, in Initial Rept. DSDP, edited by Kagami, and Karig, pp.611-632, 1986.

Tamaki, K., E. Inoue, M. Yuasa, M. Tanahashi, and E. Honza, On the possibility of Quaternary backarc spreading of the Bonin arc, Chikyu, 3, 421-431, 1981.

Taylor, B., and N. C. Smoot, Morphology of Bonin forearc submarine canyons, Geology, 12, 724-727, 1984.

Truchan, M., and R. L. Larson, Tectonic linearment on the Cocos plate, Earth Planet. Sci. Lett., 17, 426-432, 1973.

Utsu, T., Seismology, 2nd edition, 310 pp., Kyoritsu-Shuppan, Tokyo, 1984.

Yoshii, T., Compilation of geophysical data around the Japan islands (I), Bull. Earthqu. Res. Inst., Univ. Tokyo, 54, 75-117, 1979.

Yuasa, M., Geological Map of the northeast of Hachijojima, 1/200,000, Marine Geol. Map Ser., 26, 1985.

THREE-DIMENSIONAL NUMERICAL ANALYSIS OF CONTINENTAL MARGINAL BASIN DEFORMATION RELATED TO LARGE EARTHQUAKE DEVELOPMENT

Huan-Yen Loo [1]

Institute of Geology, State Seismological Bureau, Beijing, China

Abstract. The spatial relationship between intra-continental earthquake distribution and vertical deformation is discussed qualitatively. Convergent plate motion, such as the India and China plate collision, causes regional uplift with a compressive stick-slip focal mechanism, but divergent plate motion generates a regional and extensional subsidence with a focal mechanism of both strike and dip slips. Thus the long-term phenomena of macroscopic deformation reflects the characteristics of regional tectonic driving mechanisms related to the long-term prediction of earthquakes occurring either in surrounding active basins or along active strike-slip faults. Quantitatively, three-dimensional numerical analysis was carried out to match the observed surface deformation and the recent large earthquake activity surrounding the North China bay area. First, the crustal temperature distribution was calculated and used to determine the mechanical parameters of both individual layers and fault zones of the established geological model. Then, in order to satisfy the field conditions of subsidence with normal faults, the three-dimensional relationship of crustal stress and strain was computed elastically and repeatedly, using different ratios of vertical and horizontal boundary forces. It was found that, in addition to the body force, the uplift force, which is larger than the horizontal ones, should be applied immediately below the crust-mantle doming zones. Such an upwarping forces seems to result from the secondary convection which occurs in the transition zone between continent and ocean during the waning period of the Pacific plate movement. On the other hand, the local deformation near a particular fault reflects both the shallow and the deep processes. An elasto-visco-plastic model, consisting of a visco-elastic upper crust overlying a visco-plastic lower layer, is required. Model results show that (1) the transition zone between graben and horst determines the location of large earthquakes, most of which occur at the ends of active grabens, and (2) short wavelength aseismic subsidence on a non-uniform fault zone with weak asperities is due both to the cohesion loss of the fault interface and the reduction of shear modulus of the gouge of the overstressed segments, and to the effect of the (self-gravitating) overburden pressure. The increased stress, however, is due to the stress diffusion caused by the plastic relaxation of the underlying lower crust and/or upper mantle. Such an accelerated deformation, restricted to a particular fault without preshocks, may behave as a mid-term precursor. A third result is that local subsidence speeds up further, changing from short to longer wavelengths, and then it slows down suddenly. This means that the fault zone has been weakened completely and the creep of the low velocity zone below the fault has been accelerated, shedding loads into the front of the fault tip, bringing its stress to rupture level. Accordingly, the stress rate slows down, bringing about a decrease in the strain rate (i.e. energy loss in linking micro-cracks). The widening of the local subsidence zone, and the sudden stopping of high-speed strain, may represent a short-term precursor. The significant relative coseismic and postseismic subsidence along the newly faulted belt results from a combination of the elastic strain rebound during focal fracture, and of the later stage visco-elastic rebound, coupled with the action of overburden pressure. These can be utilized to predict aftershocks and impending earthquakes.

Introduction

Plate tectonics, which considers that the lithosphere is divided into a number of rigid plates interacting only near their common boundaries, can explain many of the global features of the earth. There exist, however, large continental regions where the lithosphere undergoes large-scale deformation and a rather diffuse seismicity

---

[1] Visiting Scientist, Earth Resources Laboratory, Department of Earth, Atmospheric, and Planetary Sciences, Massachusetts Institute of Technology, Cambridge, Massachusetts 02139

Copyright 1987 by the American Geophysical Union.

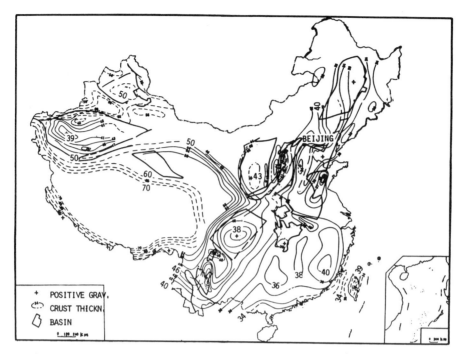

Fig. 1. Crust thickness and basin distribution of China continent.

prevails. China is one of the most interesting examples of intra-continental deformation. There are convergent collisions between plates, causing uplift in the western part of China, and divergent relaxations, causing subsidence in the eastern part of the country (Figures 1 and 2).

The crustal thickening of about 40 km is localized under the Himalayas and in the Tibetan plateau, and presents several NS oriented grabens [Molnar and Tapponnier, 1978]. This is probably due to isostatic adjustments causing E and SE oriented upper crustal creep. The crustal thinning reaches about 10 km under Shansi and Bohai bay, presenting NE and NNE oriented grabens [Ma et al., 1982], probably due to a lower eastward crustal creep which causes an oceanward migration of the Pacific Plate boundary. There is also a large horizontal strain. To the East are right-lateral strike-slip faults whose total slip is roughly of the order of a few hundred kilometers. To the West is a left-lateral fault system whose accumulated slip is a few tens of kilometers.

Generally, a horizontally oriented, maximum compressive principal stress would correspond to either thrust or strike-slip deformation and the crustal normal faulting is tightly related to the deep seated source. Therefore, the evolution of basins on the continental margin is chiefly associated with the activity of lithospheric uplift preceding thinning (Figure 3), and its cooling

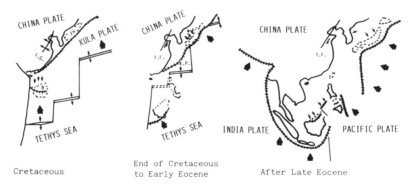

Fig. 2. Evolution of the plates around China through time (after Ben-Avraham,1978 and modified). Arrows and half arrows show relative motion of plate and blocks. Double line - spreading center, straight line - transform fault, solid line with short bars - subduction zone. T.G.= Tan-Lu graben. JA = Japan. B = Borneo. T.F.= Tan-Lu fault zone, K.P.= Kula plate.

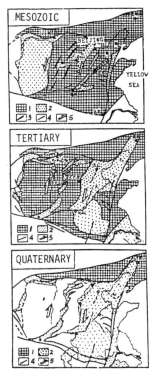

1 - UPLIFT
2 - DEPRESSION
3 - FOLDING AXIS
4 - FAULT
5 - BLOCK MOVEMENT DIRECTION

Fig. 3. Evolution of tectonic basins.

preceding subsidence. In other words, the vertical movement can be regarded as a driving mechanism in the formation of basins. These tectonic features are coupled with a regular elevation of the present-day topography: the mean elevation diminishes from the Tibetan plateau toward the Pacific Ocean.

The analysis of temporal and spatial changes in deposits of the North-China Plain would furnish some clue for the study of these phenomena. Based on the study of typhonic inclusions in the basaltic rock, Zheng [1980] pointed out that the magmatic activity in this plain behaved as an intense fissure erruption, spreading over the whole region in Early Tertiary, but reduced to local erruption during the Quaternary, and that the formation temperature of magma was 1100-1300° C. But the magnetotelluric soundings made recently by Liu and others [1982] illustrate that the lithosphere thickness remains the same. All these phenomena show that the lithosphere has established a new equilibrium thickness. Thus the cooling and thickening processes, after an intensive thermal event lasting about 50 m.y., agree with the lithospheric thermal model of the Central North Sea Basin. Obviously, there is a relationship between the mean surface heat flow in a continental area and time, if that area has experienced a thermal event. However, the continuous subsidence, average heat flow value > 2.0 and large earthquakes occurring in and surrounding the North-China Plain, raise a question about the mechanism. One would

have to recognize that the induced convection [Toksoz et al., 1978] in the mantle by subduction of the Pacific plate may lead to a lateral density contrast causing possibly a small secondary convection of the deep seated materials beneath the North-China bay area, and then a change of the thermal regime and structure, causing, in turn, an isostatic adjustment, and becoming the driving force which generates large crustal earthquakes and results in higher heat flow. As the deep source (i.e. the induced convection) wanes, the crustal stress redistribution takes place and is related to earthquake development [Loo, 1883].

Evidently, the regional tectonic phenomena cannot be isolated but must be analyzed from various aspects. For instance, the high temperature creep of deep seated materials affects, perhaps, the regime of shallow materials. To determine how to use the observed or explored geological and geodetic data, estimating the situations below approximately, would be a prerequisite to quantitative investigation in earthquake prediction.

One of the major limitations of the earthquakes study in the North-China Plain, is their infrequent occurrence. Conventional information, such as spatial and temporal trends in seismicity, focal mechanisms, surface faulting, etc., are generally lacking. Because leveling information must still be regarded as a potentially valuable source of intra-continental crust dynamics information, leveling results in such an area were used recently to study intraplate deformation and seismicity in spite of the uncertainty caused by systematic leveling errors and near surface effects.

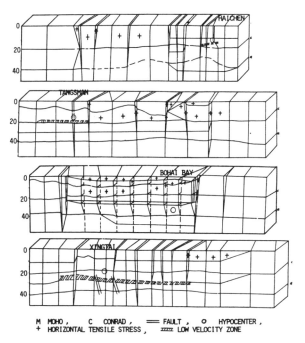

Fig. 4. Three dimensional numerical modeling of North China plain.

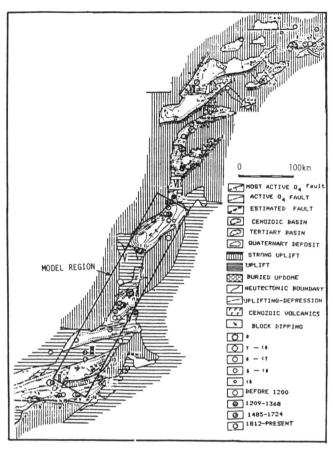

Fig. 5. Seismo-geological map of Shanxi graven system.

The time behavior of recent crustal movement is an important factor for deducing their origin and their possible implication for earthquake predictions. Some seismically active areas suggest rapid rate changes in deformation. Of course, rapid crustal deformation is not necessarily indicative of an impending earthquake. Even along major plate boundaries, where earthquakes occur relatively frequently, and the basic processes responsible for their occurrence are generally understood, the nature of inter-seismic and possibly pre-seismic movements remains uncertain. The problem is much more complicated in intra-continents, where earthquakes are rare and their causes are generally unknown.

In spite of the substantial difficulties surrounding intraplate earthquakes, leveling measurements have proven effective for monitoring deformation associated with earthquakes located away from major plate boundaries. Releveling measurements continue to be the most accurate and widespread source of information on vertical movements of the continental crust. They constitute an important input to earthquake prediction. However, releveling estimates of crustal movement are influenced by near surface movements and suffer from poorly understood systematic errors which can be mistaken for tectonic deformation. Integrating other geophysical and geological information with leveling results is an essential element for proper interpretation.

All the recent large earthquakes in the North-China Plain are associated with dip-slip fault movement. There is no clear evidence for permanent vertical deformation associated with purely strike-slip faulting, although present observations are not sufficient to rule it out. Several investigators [Jurkowski, et al., 1981] have also reported deformation possibly associated with impending earthquakes or aseismic fault movements. Co-seismic and post-seismic movements are well established for larger (M>6) dip-slip earthquakes, but evidence for pre-seismic deformation is generally lacking or ambiguous. Relating observed deformation to preseismic mechanisms may be quite difficult because of the limited understanding of precursory phenomena, and of the inability to distinguish them from vertical movements due to other causes, such as magmatic activity, isostatic movement, and orogenic deformation. Profile examination does not always clearly indicate more regional trends of elevation change. Mapping is also an important tool for examining vertical crustal movements. Movements with lateral continuity can be identified and compared with areal geological and geophysical observations. A consistent feature displayed by the North-China Plain leveling data (Figures 7 and 8) is the oceanward tilting of the plain. The pattern of Cenozoic sedimentation is consistent in signs with the geodetic observations.

In Figure 9, the anomalous short and longer wavelengths of the observed subsidence, as well as the sharp change in the relative deformation of the coseismic fault, are neither the measured errors, nor the compaction of dewatered sediments and should reflect both the shallow and deeper processes of the lithosphere. Why and how four large earthquakes (M>7.0) occurred in the North-China Plain area within 10 years and why only one of these shocks had intensive pre-microshocks

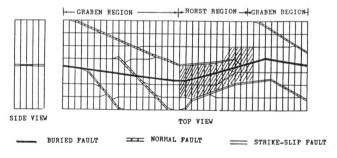

Fig. 6. Three dimensional model of Shanxi graven system.

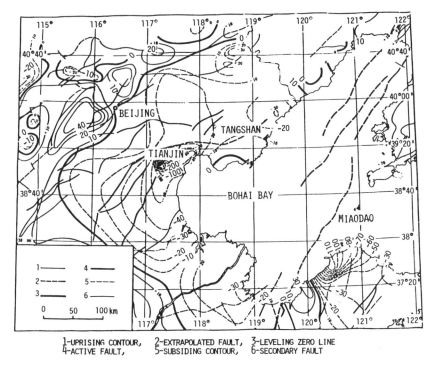

1-UPRISING CONTOUR, 2-EXTRAPOLATED FAULT, 3-LEVELING ZERO LINE
4-ACTIVE FAULT, 5-SUBSIDING CONTOUR, 6-SECONDARY FAULT

Fig. 7. Leveling contours of 1972-1953 in North China plain.

remains one of the fundamental unsolved problems of earth sciences in the plate tectonic era.

The model we propose to investigate in detail depends on the strength of the asperities in the pre-existing fault. If the initial rupture of the earthquake occurs at highly-stressed asperity in the fault, there would be short term precursors. If the rupture (shear sliding) starts in a low-

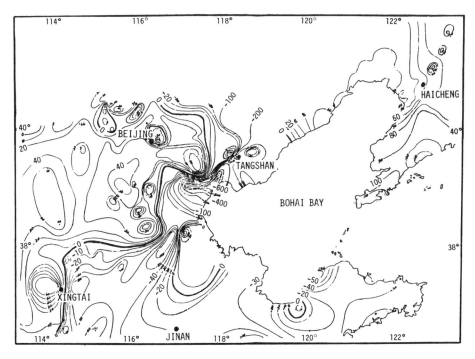

Fig. 8. Leveling contours of 1966-1979 in North China plain.

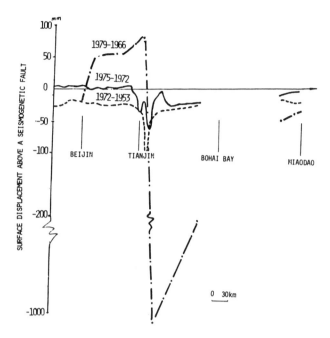

Fig. 9. A profile crossing Bohai Bay area northwesterly.

stress region and without strong asperities inside the pre-existing fault, then there may not be any observable anomalies in the epicentral region, except aseismic fault creeping. The magnitude may be estimated from the size of the seismic gap and the area of the deformed region. The time prediction for the occurrence (in weeks, days or hours) depends on short-term precursors. However, based on the asperities and stress distribution on the fault, the precursors may or may not be present. For faults with large areas and many asperities, especially for many strike-slips associated with dip-slip faults, it may be difficult to predict where and when the rupture may start, and once started, how far it would propagate. Similarly, it may also be difficult to determine whether the accumulated stress will be released as one large earthquake or as a sequence of small events. This results in greater uncertainty about the magnitude, as well as the time, of occurrence [M.N. Toksoz, 1983, personal communication]. The problem of computing the surface displacement due to various internal sources, including the earthquake induced ground deformations, received considerable attention over the years. However, the time-dependent ground deformation is a superimposed phenomena of regional isostatic adjustment, localized aseismic fault slip, and hypocenter strain accumulation. The measured surface movements reflect complex processes in the lithosphere. How to distinguish them, and how to incorporate the regional long-term, slow deformation, as well as the local short-term, fast deformation into a numerical model is a difficult but necessary task. This paper looks at geodetic data (especially leveling) from the North-China Plain, relates it to earthquake development, and creates a 3-dimensional, numerical model to learn the behavior of a strike and dip-slip fault.

Most previous work modeled only deformations which follow closely an earthquake in time but neglected gravitational effects. However, many applications, such as earthquake cycle and cyclic faulting, require calculations for long periods of time, hence, gravity must be included in the models [Savage, 1983]. Our model will include source function at depth, temperature-dependent layered earth structure, regional tectonic stress regime, and the gravitational effects.

Finite Element Analysis and Discussion

A number of authors [Vilotte et al., 1982] have used 2-dimensional finite element techniques in trying to model lithospheric deformation, neglecting the gravitational effect, and treating the crust either as a free body (i.e. plane stress), or a rigidly and vertically constrained body (i.e. plane strain). None of these can reflect the reality of continental tectonics. We propose an elasto-visco-plastic 3-dimensional numerical model for both long-term and short-term deformation, depending upon a the crustal temperature distribution and driving mechanism. Theoretical models for earthquake instability and precursory deformation seem to be now developed well enough to justify earthquake prediction. Instability models generally assume a single fault, and are represented as a suddenly imposed dislocation in the source [Rice, 1983, Rundle, 1980, Savage and Thatcher, 1983] a fact which cannot explain the post- and pre-seismic deformation. Actually, a fault zone in the field is an irregular body compiled of smaller, localized zones enclosing fragments. In addition, the earth contains numerous interacting faults driven by poorly understood thermo-mechanical processes.

This study requires an approximation of the average behavior of the lithosphere. For a short period of time, the upper crust behaves visco-elastically or elastically while, for longer peri-

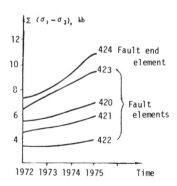

Fig. 10. Stress build-up at end of seismogenic fault.

ods of time, it creeps non-linearly. The lower crustal deformation is controlled by temperature-dependent, non-linear creep law as a visco-plastic medium. In order to match the observed surface deformation and the recent large earthquake activity surrounding the North-China bay area, the crustal temperature distribution is calculated [Gao, 1983] and then used to determine the mechanical parameters of both individual layers and fault zones of the established geological model composed of sedimentary, granitic, basaltic, and upper mantle layers, The proper choice of boundary conditions is ambiguous. The three-dimensional relationship of crustal stress and strain is then computed elastically and repeatedly, using different ratios of vertical and horizontal boundary forces to satisfy the field conditions of regional subsidence with normal faults. It is found (Figure 4) that, in addition to the body force, the uplift force, larger than the horizontal ones, would be applied intermittently below the crust mantle doming zone [Loo et al., 1983]. On the other hand, the local deformation near a particular fault reflects both the shallow and the deep processes. An elasto-visco-plastic model, consisting of a visco-elastic upper crust overlying a visco-plastic lower layer, is required. Model results show that (1) the transition zone between graben and horst determines the location of large earthquakes occurring mostly at the ends of active gravens (Figures 5 and 6); (2) aseismic subsidence of short wavelength (Figure 9), on a non-uniform fault zone with weak asperities, is due to, both the cohesion loss of the fault interface and the shear modulus reduction of the gouge of the overstressed sediments, and to the effect of the overburden pressure (self-gravitating) effect but that the increased stress comes from the stress diffusion of the plastic relaxation of the underlying lower crust and/or upper mantle. Such an accelerated deformation, restricted to a particular fault

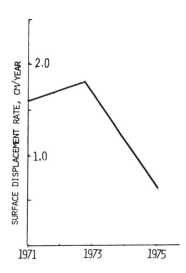

Fig. 12. Surface displacement rate over seismogenic fault.

without preshocks, may behave as mid-term precursor; (3) local subsidence, speeding up further, changing from short wave to longer wavelength, and slowing down suddenly (Figures 11 & 12), means that the fault zone has been weakened completely and the creep of the underlying low velocity zone has been accelerated, both shedding load onto the front of the fault tip, bringing its stress to rupture level and thus decreasing the strain rate (due to the stressing rate slowing down, i.e. energy loss in linking micro-cracks). The widening of the local subsidence zone, and the sudden stopping of the high speed strain, may represent the short-term precursor; and (4) the significant relative co-seismic and post-seismic subsidence along the newly faulted belt is the combination of the elastic strain rebound during focal fracture and the visco-elastic rebound in the later stage, together with the action of overburden pressure. These can be utilized to predict aftershocks and impending earthquakes.

Conclusion

The available theoretical models for studying strike-slip or dip-slip do not always display the same features because of the different mathematical approaches involved. Only 3-dimensional models can simultaneously model both types of motion and are most appropriate to fit the observations.

The convergent plate motion, such as the India and China plate collision, causes regional uplift with a compressive stick-slip focal mechanism, but the divergent plate motion generates a regional and extensional subsidence with a focal mechanism of both strike and dip slips. Thus the long-term phenomena of macroscopic deformation reflect the characteristics of regional tectonic driving mechanisms related to the long-term prediction of

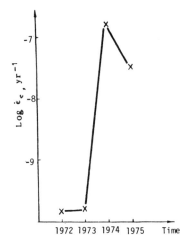

Fig. 11. Strain rate speed-up at seismogenic fault end.

earthquakes occurring either surrounding active basins or along active strike-slip faults.

Shallow crustal seismicity and stress states are influenced by the change in the aseismic creep rate. The localized and sharp aseismic dip-slip, on a non-uniform fault without pre-microshocks, is an indication of the weakening of the pre-existing fault zone, and may provide a plausible explanation for features associated with seismicity preceding a major event. It indicates, probably, a mid-term precursor. The accelerated and widened aseismic slip event may be an indication of the plastic weakening of the deep material below the reactivated fault-tip and may be regarded as a short-term precursor. All these events would cause stress increase on the locked and unfailed portion in the front of the fault tip, bringing it to rupture level and generating a great earthquake.

However, areal coverage of geodetic measurements is generally incomplete and the precise mechanisms responsible for the observed deformation are uncertain. Caution is needed when applying tectonic interpretation to releveling observations.

Acknowledgements. The author gratefully acknowledges the financial support from the Department of Earth, Atmospheric, and Planetary Sciences under the NASA grants (NAG 5-41 and NAG 5-44) to write this paper while visiting MIT, and appreciates Professor M. N. Toksoz and Professor T.R. Madden for their comments and suggestions on this paper.

References

Dimowska, R. and J.R. Rice, Fracture theory and its seismological applications, Div. of Applied Sci., Harvard Univ. 1983.

Gao, X.L. and H.Y. Loo, Three-dimensional modeling of lithosphere temperature distribution in North China Bay area, Institute of Geology, State Seismological Bureau, Beijing, China, 1982.

Jurkowski, G. and R. Reilinger, Recent vertical crustal movements: The eastern United States, Dept. of Geological Sci., Cornell Univ. 1981.

Lehner, F.K., V.C. Li and J.R. Rice, Stress diffusion along rupturing plate boundaries, J. Geophys. Res., 86, 6155-69, 1981.

Li, V.C. and J.R. Rice, Precursory surface deformation in great plate boundary earthquake sequences, Bull. Seism. Soc. Amer., 73, 1415-34, 1983.

Liu, G.D., The structure of the crust and mantle in the north of North China and its relation of the Cenozoic tectonism, Institute of Geology, State Seismological Bureau, China, 1982.

Loo, H.Y. and H.Z. Song, Three-dimensional modeling of large earthquakes migration, Institute of Geology, State Seismological Bureau, Beijing, China, 1982.

Loo, H.Y. and X.S. Jin, Finite element analysis for earthquake development with ground deformation, Symposium of the recent plate movements and deformation, Tokyo, Japan, 1982.

Loo, H.Y. and W.A. Gao, Numerical modeling of formation mechanism of Shanxi Graben of North China, IUGG, Quantitative Geodynamics Symposium, Hamburg, W. Germany, 1983.

Ma, X.Y., Q.D. Deng and Y.P. Wang, Cenozoic Graben systems in North China, Z. Geomorph. N.F., Suppl. Bd., 42, 99-110, 1982.

Michael, J.R. and R. Reilinger, Land subsidence due to water withdrawal in the vicinity of Pecos, Texas, Eng. Geology, 11, 295-304.

Molnar, P. and P. Tapponnier, Cenozoic tectonics of Tibet, J. Geophys. Res., 83, 5361-75, 1978.

Rice, J.R. and J.C. Gu, Earthquake aftereffects and triggered seismic phenomena, Div. of Applied Sci., Harvard Univ., 1983.

Reilinger, R., Elevation changes near the San Gabriel Fault, Southern California, Geophys. Res. Letters, 7, 1017-19, 1980.

Rundle, J.B., Static-elastic-gravitational deformation of a layered half space by point couple sources, J. Geophys. Res., 85, 5355-63, 1980.

Savage, J.C., A dislocation model of strain accumulation and release at a subduction zone, J. Geophys. Res., 88, 4984-96, 1983.

Stuart, W.D., Forecasting earthquake instability with geodetic data, Solid earth geophysics and geotechnology, AMD - Vol. 42, 1979.

Thatcher, W., A viscoelastic coupling model for the cyclic deformation due to periodically repeated earthquakes at subduction zones, U.S.G.S., Menlo Park, CA, and Sandia Lab., Albuquerque, New Mexico, 1983.

Toksoz, M.N. and A. Hsui, Numerical studies of back-arc convection and the formation of marginal basins, Tectonophysics, 50, 177-96, 1978.

Vilotte, J.P., M. Daignieres and R. Madaraga, Numerical modeling of intraplate deformations: Simple mechanical model of continental collision, J. Geophys. Res., 87, 1709-728, 1982.

Zheng, X.Z., Some deep-seated inclusions in basalt from the North China fault block region and problems about basalt, Formation and development of the North China block fault region, 335-347, Science Press, China, 1980.

NEOTECTONIC DEFORMATION OF THE ALPIDE FOLD BELT IN THE CENTRAL
AND EASTERN MEDITERRANEAN AND NEIGHBOURING REGIONS

N. Pavoni

Institut für Geophysik, ETH-Hönggerberg, Zürich, Switzerland

Abstract. The Alpide fold system of the Mediterranean region represents a strongly mobilized zone of crustal deformation between the more stable North-European and African platforms. Neotectonic deformation is present everywhere within the fold belt. There are no microplates within the belt. Kinematic analyses of neotectonic structural features reveal some large-scale regularities in the pattern of neotectonic deformation which are briefly presented and discussed.

Introduction

The map shown in Figure 1 is an attempt to illustrate the pattern of neotectonic deformation within the Alpide fold belt in the central and eastern Mediterranean region. The map is based on data derived from numerous neotectonic structures, i.e. structures which were formed or reactivated during the Pliocene or the Quaternary.

Construction of the Map

The structural features used in the analysis include:
a) folds and thrust faults,
b) normal faults and graben structures,
c) strike-slip faults and other faults with strong component of lateral displacement (see maps in the references, and also, national geological, tectonic and seismotectonic maps of Mediterranean and neighbouring countries).

The orientation of maximum horizontal crustal shortening (MHS) was determined from the trends of fold axes and the strikes of reverse and strike-slip faults. Maximum horizontal shortening is assumed to be oriented orthogonally or at a high angle to the trend of fold axes and to the strike of reverse faults. In the case of strike-slip faulting interpreted as shear faulting, the orientation of MHS is at an acute angle of varying amount - but usually near 45° - to the strike of the fault.

The orientation of maximum horizontal crustal

Copyright 1987 by the American Geophysical Union.

extension (MHE) was derived from normal faults and graben structures. MHE is assumed to be oriented orthogonally or at a high angle to the strike of normal faults. In the case of strike-slip faults, MHE was usually taken at an angle of about 45° to the strike of the fault.

The orientations of principal horizontal strain derived from local neotectonic structural features as well as the orientations of shear strain given by the strike-slip faults constitute the basic data of the map (Figure 1, legend 1-4). To better represent the general strain field, trajectories of MHS and MHE were constructed (Figure 1, legend 5-6).

Discussion

In discussing the map the following points have to be considered: the map is based on near surface deformation data; only the horizontal component of deformation is considered; the map is a twodimensional, qualitative representation of neotectonic strain; and the tectonic structures and evolution of the region covered by the map are of great variety and complexity.

Despite these limitations, several observations regarding the neotectonic deformation of the eastern Mediterranean region are of interest:

(1) Neotectonic deformation is present everywhere within the fold belt. There are no "rigid" segments of the crust (microplates) within the fold belt.

(2) In certain regions the neotectonic strain field displays a rather regular pattern. The MHS trajectories appear as straight lines trending parallel to each other over large distances, as do the MHE trajectories. A regular NNW-SSE trend of MHS trajectories is observed in the folded zone of Tunisia and eastern Algeria. The same trend continues towards the west into the Atlas Mountains of Algeria and Morocco. There are indications that the same trend of MHS trajectories extends into the northern part of the Sahara platform.

A regular pattern of the strain field is observed in the Dinarides belt and the Adriatic region with NE-SW trending MHS trajectories.

In the Central and Eastern Alps, the MHS traj-

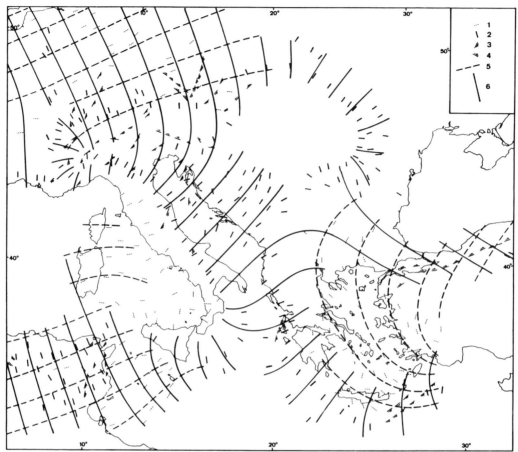

Fig. 1. Neotectonic deformation of the Alpide fold belt in the central and eastern Mediterranean regions. Legend: 1 – Orientation of maximum horizontal crustal extension (MHE); 2 – Orientation of maximum horizontal crustal shortening (MHS); 3, 4 – Lateral shearing; 5 – MHE trajectories; 6 – MHS trajectories.

ectories are oriented NNW-SSE. It is notable that the same NNW-SSE orientation of MHS is observed in Central Europe in the foreland of the Alps. MHE is oriented ENE-WSW, as documented by Quaternary faulting in the Upper Rhinegraben area, in the Lower Rhenish Embayment and Czechoslovakia. The MHS orientation does not change when proceeding from the interior of the Eastern Alps to the northern foreland. Without doubt, the rates and amounts of deformation within the fold belt have been much larger than in the foreland.

(3) In certain segments of the Alpide fold belt, as e.g. in the Western Alps, the Carpathian Mountains and the Aegean region, a distinct fan-like pattern of MHS trajectories can be observed. These fan patterns clearly correspond to the arcuate shape of these mountain belts. In general, the MHS trajectories are oriented perpendicularly, the MHE trajectories more or less parallel to the strike of the arc.

Arc formation is characteristic for Alpide fold mountains. In the case of the Western Alps, for example, it can be shown that the orogenic movements and tectonic transport progressed from the internal parts to the external parts of the arc. Early compressive phases occurred in the Piemontese zone and later phases successively proceeded into the Brianconnais and Dauphinois zones. The youngest compressive movements (Upper Tertiary) are manifest in the fold belt of the Jura Mountains and the Chaines Subalpines. The orientations of MHS as derived from kinematic analyses of Miocene and Pliocene folds and faults are distinctly fan-like and at high angles to the trend of the arc (Figure 2). The map represents only a two-dimensional picture of the tectonic strain field. For a thorough understanding of arc formation, more information about the deformation in the third dimension, i.e. at greater depth, would be needed.

In the back-arc region of the Aegean arc, i.e. in the region of the Aegean Sea, and in the back-

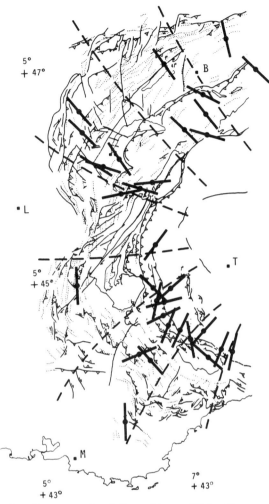

Fig. 2. Neotectonic deformation within the arc of the Western Alps. Dashed lines: MHS trajectories as derived from kinematic analyses of neotectonic structural features. Bars: Orientation of P-axes from fault-plane solutions of earthquakes. B: Bern; L: Lyon; M: Marseille; T: Torino.

arc region of the Calabrian arc, i.e. in the region of the Tyrrhenian Sea, extensional tectonic regimes are observed.

(4) A systematic investigation of focal mechanisms of earthquakes yields valuable information about the present tectonic strain and stress states of the crustal segments in which the earthquakes occur. It is of special interest to compare, region by region, the results of seismological research with the results of kinematic analyses of neotectonic structures. In many regions, such as the Western Alps (Figure 2), Switzerland, Eastern Alps, Aegean region, western Turkey and North Africa, the orientations of P- and T-axes derived from fault-plane solutions of earthquakes correspond well with the orientations of MHS and MHE trajectories shown in Figure 1. Evidently, the stress field which causes the present-day seismicity in these regions is very similar in its orientation to the stress field which, 5-10 million years ago, produced the neotectonic deformation. In other regions, such a continuity between present-day and neotectonic stress and strain fields is not observed.

(5) The NNW-SSE orientation of MHS trajectories and the corresponding ENE-WSW orientation of MHE trajectories which are observed in Central Europe north of the Alps, in the Central and Eastern Alps, as well as in the young fold belt of North Africa, are an expression of the N-S to NW-SE oriented convergence of the European and African plates.

This investigation will be extended to the Near East and to the western Mediterranean region and should provide useful information in reconstructing the evolution of the fold belt.

References

Carte tectonique internationale de l'Europe, 1:2 500 000, N. Schatsky et al., eds., 1962.
Carte de la tectonique actuelle et recente du domaine Mediterraneen et de la chaine Alpine, 1:2 500 000, H. Philip, ed., 1982.

THE STUDY OF STRESS AND STRAIN INHOMOGENEITIES AT VARIOUS SCALES IN USSR

G. A. Sobolev

Institute of the Physics of the Earth, Academy of Sciences, Moscow, USSR

Abstract. The paper reviews recent work of Soviet scientists related to stress and strain measurements by geophysical methods both in the laboratory and in situ. Laboratory experiments on deformation of large rock blocks may reveal fine inhomogeneities of their stress and strain field. They are only few centimeters in size, and could not be recognized, before loading, visually or by ultrasonic and electrical sounding. Seismic investigations in a broad frequency range reveal the inhomogeneities of a larger size, from a few centimeters up to a hundred, which are connected with the block structure of rock massif. Seismotectonic strain measurements tell us about the mosaic space distribution of strain field on the scale of kilometers. A detailed study of strain distribution was made on seismoactive region in Central Tajikistan. The regularity of subsequent block sizes in rock masses was revealed by summarizing the geological, geodetic and geophysical data. The ratios between each ensuing and preceding size equal approximately 3 and manifest themselves for length-scales by the factor of 10. Establishment of our knowledge in a broad scale range about block structures of rocks and the related inhomogeneities of stress-strain field in various regions is an important problem. It is necessary, for this purpose, to combine all the methods permitting integration of the medium properties of various sizes of rock masses.

Introduction

This report reviews the recent work of Soviet scientists related to stress and strain measurements, both in a laboratory and in situ, with emphasis on results showing inhomogeneity of the stress and strain field in rocks. The inhomogeneity can substantially impede interpretation of geodetic and seismological data and requires employing a dense network of stations. As this inhomogeneity appears in various scales, it should be investigated by all the available methods which permit the integration of the medium properties at various measuring distances. Incorporation with seismological studies of these inhomogeneities tells us about the block structure of the Earth's rocks together with the related distribution of their stress-strain fields. Without this sort of knowledge, for example, one cannot reliably predict the largest size of earthquakes which will occur in each region.

Results of Laboratory Measurements

A possible inhomogeneity of the stress-strain state of rock due to its granular structure is obvious and well known. Here, the scale of the relatively homogeneous zones is only a few millimeters. Experiments on deformation and destruction of large blocks with a side length of ca. 1 m were conducted in the Soviet Union using a 50,000 ton press and revealed inhomogeneities on the scale of a few centimeters [Sobolev et al., 1982]. It is notable that they could reliably record the inhomogeneity even at early stage of loading (under 30% of the breaking load). However, the rock structure of such a small size could not be seen before the loading visually or by ultrasonic or electrical sounding.

To record the strain field, a large number (ca. 100) of strain gauges were bonded to the surfaces of granite and basalt blocks. The gauges were placed vertically (parallel to the load application axis), horizontally and at 45° to the axis. The readings of the gauges were used to calculate all the components of the surface strain. Figure 1 shows isoline maps of the first invariant of the plane strain tensor, $I = \varepsilon_v + \varepsilon_h$, for the central part of the granite block surface, where ε_v and ε_h are the vertical and horizontal strain components, respectively. The granite block was a cube of 70 cm in side length. The maps in Figures 1a and b relate to the load, (a) 113 MPa and (b) 134 MPa, which correspond to 80% and 96% of the breaking load, respectively.

Analysis of over 100 of such maps for each sample, obtained at different stages of loading, demonstrates that all the samples are characterized by mosaic patterns of zones of varying contraction. Their distribution over the surfaces shows a very slow change in time, remaining intact up to the block destruction, whereas their

Copyright 1987 by the American Geophysical Union.

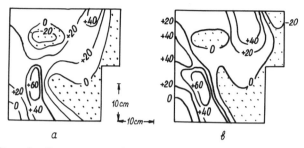

Fig. 1. Structures of the strain field $I = \varepsilon_v + \varepsilon_h$ for the granite sample at 0.8 (left hand side) and 0.96 (right hand side) of the breaking load. Isolines are drawn through $I = 20 \cdot 10^{-5}$ units. Dilatancy regions are indicated by dots.

absolute value changes significantly. Sometimes the lowest contraction zones change to dilatant zones. Tensile fractures are primarily observed in the regions of large dilatancy whereas the shear fractures are typical of zones with anomalously large shear strain.

It should be noted that inhomogeneous zones of the same size, i.e. a few centimeters, are also revealed on the blocks by ultrasonic testing. The structure manifests itself by zones of a simultaneous increase and decrease of the elastic wave velocities. Note that the mosaic picture of the strain field in the central parts of the blocks cannot be attributed to the stress concentration as seen on their corners or surfaces under the effect of the press. Thus its origin appears to be due to the inherent structure of the studied rocks themselves.

Study of Rock Massifs

A similarly complex picture of stress-strain fields can be observed under the natural conditions, particularly in the studies of large rock masses during construction of hydroprojects. A method of determining rock stresses from seismic data has been developed elsewhere [Savich and Yashchenko, 1979; Savich et al., 1981]. This method consists of measurement of seismic wave velocities at different frequencies, together with calibration of the velocity as a function of the applied stress when the rocks are unloaded as a result of the underground works. This method is advantageous in comparison with other stress measurements by hydrofracturing or coring, as it permits determination of the mean stresses in blocks of varying sizes.

The parameters of the rock volume W, which are subjected to integration, are related to the wavelength λ or the elastic wave frequency f, wave velocity v and measuring distance l:

$$W = \pi(\alpha\lambda)^2 l \approx 0.2\frac{lv^2}{f^2}, \quad (1)$$

where $\alpha \approx 0.25$ is the wave penetration index and $\pi = 3.14$ [Savich, and Yashchenko, 1979].

Figure 2 is a section across the valley of the Inguri river with the marked values of the horizontal component σ_y along the valley [Savich et al., 1981]. The stress was estimated by the abovementioned seismic method and by direct measurements in the holes. The massif is composed of limestones cut with a system of tectonic fractures whose major types are shown in Figure 2. It is evident that noticeable variations in the stress fields are recorded due to natural weakening near the tectonic disturbances. The remaining stress components σ_x and σ_z vary in the same manner although their absolute values are smaller.

The stresses decrease with measuring distance l. When passing from $W = 10^{-3}$-10^{-2} m^3 to $W = 10$-10^2 m^3, the stresses in the region of the Inguri and Toctogul hydroelectric stations drop for ca. 50% and 25%, respectively. The inhomogeneity of the stress field structure is closely associated with the block structure of the massif with the typical blocks sized 0.6 m, 2.1 m, 9 m, 50 m and 200 m, according to field geological observations. G.A. Markov [1983] reported extremely high horizontal stress in the Earth's crust as result of upward tectonic movements. The magnitude of the stress near the Earth's surface reaches several

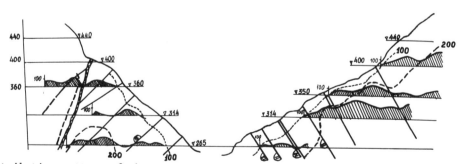

Fig. 2. Distribution pattern of the stresses σ_y at the Ingury hydroelectric station (section along the drainage galleries). The shaded zone shows a level of σ_y along the openings. The long and short dashes indicate the isolines $\sigma_y = 200 \cdot 10^5$ N/m^2 and $100 \cdot 10^5$ N/m^2, respectively. Solid lines indicate tectonic fractures.

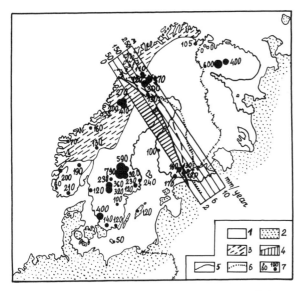

Fig. 3. The stress distribution as measured in the mines of Scandinavian peninsula.
Rocks: 1- Precambrian, 2- Paleozoic, 3- Caledonian. Profiles: 4- uplift during last 7.5 years (m); 5- recent vertical movements (mm/year); 6- post-glacial uplift(m); 7- stress magnitude; the weight of overburden is excluded (bars).

hundred bars in the regions of uplifting shields. Figure 3 demonstrates the maximum horizontal stress distribution on Scandinavian peninsula based on overcore data [Markov, 1983]. The largest stresses were registered in the central parts of recent and post-glacial uplifts.

The existence of locally high stresses was explained by V.S. Ponomarev [1981] who introduced the hypothesis of "zoning relaxation." He proposed that the stress relaxation process in an uplifting massif starts from the edges and moves inward with a velocity 10^{-5}- 10^{-6} km/year. Thus, the predicted relaxation time for central part of blocks on the scale of 10-100km is about 10^6- 10^8 years, which is significantly greater than the usually considered relaxation period 10^2- 10^4 years.

Accepting the concept of high stresses in uplifting blocks, one might expect an inhomogeneous mosaic distribution of stress-strain fields in regions undergoing vertical uplift.

Seismological Observations

In the studies of stress-strain state and block structures in the Earth's crust (scales of the order of ten kilometers), the principal technique used recently in the USSR is calculation of the seismotectonic strain. Part of the tectonic strain caused by seismogenic faults can be attributed to the volume W in which the faults occur. Here we assume that each fault length is far smaller than the size of the volume under study. The rate of seismotectonic strain ε_{ij} in a given volume can be estimated if the integration time interval T is much longer than the periodicity of the strongest (in the same volume) earthquakes [Riznichenko et al., 1982]:

$$\dot{\varepsilon}_{ij} = \frac{1}{2\pi\mu WT} \sum_{k=1}^{N} (M_{ij})_k , \qquad (2)$$

where μ is the shear modulus, $(M_{ij})_k$ are the seismic moment tensors of the k number of foci of the earthquakes in W over T.

The moment tensors have been found for only a limited number of earthquakes. Therefore the value of ε_{ij}, in practice, is calculated as a function of the earthquake energy scale, seismic activity, slopes of b-values and fault plane solution parameters related to the seismic moment tensor by correlation dependences.

Using this technique, maps of the regions of relative extension and contraction as well as shear strain have been constructed for the Pamir and Tien Shan zones at the junction of the Indian and Eurasian plates [Riznichenko et al., 1982]. Similarly, O.V. Soboleva [not published] has made a detailed investigation of the maximum extension, contraction and shear strains for the Central Tajikistan in the Soviet Middle Asia. About one thousand focal mechanisms were analysed.

Figure 4 shows an example of maps indicating a path of the maximum shear strains for two seismo-

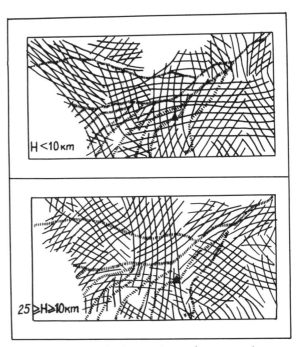

Fig. 4. Isolines of the maximum shear strains calculated from the seismotectonic strain rate for Tajikistan. Dashed lines indicate major faults. The Dushanbe station is denoted by a triangle.

genic layers of the crust in Tajikistan with $H < 10$ km and 10 km $< H < 25$ km. The studied region was divided into unit cells with lengths between $1/6°$ and $1/3°$ in area with respect to the seismic activity level.

It follows from Figure 4 that the region can be sub-divided into blocks with different orientations of the maximum shear strains. The dimensions of the blocks vary between 20 and 50 km. Note substantial discrepancies in the orientation of the isolines for the two seismogenic layers lying at different depths. Integration of the data obtained over the entire region suggests that as a whole the Central Tajikistan undergoes a nearly horizontal NW-SE compression and a sub-vertical extension (oriented NE-SW with a plunge 63°). Thus, calculations of the seismotectonic activity during the period 1955-77 indicate a vertical growth of the seismoactive layer and its horizontal contraction at a rate of $4 \cdot 10^{-9}$ year^{-1}. This value is attributed, of course, to a narrow time interval of about 20 years. At the same time, the mosaic distribution and orientation of the strain field, revealed by our method seems to be not so sensitive to the length of observation time.

Theoretical Approach

Since seismic moment is defined by the relative displacement Δu, of the rupture surfaces of area S, it is possible to calculate the Earth's strains from the seismic moment tensor. For a homogeneous elastic half-space, such a problem has been solved elsewhere [Voevoda, 1980]. Displacement, strains and tilts on the Earth's surface are functions of the coordinates of the sources (seismogenic ruptures), receivers, elasticity modulus of the medium as well as components of the seismic moment tensor:

$$M_{ij} = 2\pi \int_S \Delta u_i n_j dS \, , \quad (3)$$

where $\Delta u_i = (\Delta u_i^+ - \Delta u_i^-)$ is the relative displacement of the rupture sides, n_j is the components of the vector of the unit normal to the rupture surface S.

Rough estimations of the absolute displacements $|u|$ and strains $|\varepsilon|$ on the Earth's surface, far from the ends of the ruptures, can be made as follows:

$$|u| = (1/2\pi\mu) \sum_{k=1}^{N} M_{o(k)} / R^2_{(k)} \, ,$$

$$|\varepsilon| = (1/2\pi\mu) \sum_{k=1}^{N} M_{o(k)} / R^3_{(k)} \, . \quad (4)$$

where R is the distance between the source and the observation point, k is the number of the corresponding fault, $M_{o(k)}$ is the seismic moment of the fault. The precise equations are given elsewhere [Voevoda, 1980].

Thus, it is possible to integrate the results of the seismological and geodetic measurements in seismoactive regions. A good agreement between the strains calculated from studies of the earthquake foci and those found by direct measurements was exemplified by the Daghestan earthquake of May 14, 1970 of magnitude 6.6 [Voevoda, 1984].

Inhomogeneity of the stress-strain state is heavily influenced by the presence of ruptures in the medium. It should be noted that the stress-strain fields are distorted only by active faults, i.e. those that respond to a change of the external influence by displacement of their sides. Passive faults, particularly those unfavorably oriented towards the external fields, do not appear to affect the stress-strain fields, but these faults are also not easy to detect by elastic or electro-magnetic sounding. Transition of the rupture from the active to the passive state under a partial unloading has been demonstrated by laboratory experiments [e.g., Spetzler et al., 1981]. The same effect may be observed on the Earth's scale, too. We do not say now about the healing in a geological sense. In the present case we are concerned with a purely physical phenomenon of closing the rupture sides and restoration of the medium continuity.

The existence of active faults in a tectonic stress field leads to a situation when one region can exhibit stress or strain fields of different ranks, revealed by use of various measuring distances. The reasons for the initiation and demonstration of stress fields of different scale ranks have been discussed elsewhere [Myachkin et al., 1982; Osokina and Fridman, 1982].

Let us denote an external stress field as I-st rank one and consider formation of the II-nd rank stress field as exemplified by solution of the plane strain problem [Osokina and Fridman, 1982]. Assume that the fracture of length $L = 2 \cdot 1$ (Figure 5) is activated by a uniaxial compression I-st rank field, i.e. $\sigma_2^I = -2$, $\sigma_1^I = 0$. Then $\tau_{12} = (\sigma_2^I - \sigma_1^I)/2 = -1$, $p_{12} = (\sigma_2^I + \sigma_1^I)/2 = -1$. Let us assume a friction coefficient on the fracture tg $\phi = 0.2$, the angle between the direction σ_2^I and the fracture projection $\alpha = 45°$ and $l = 1$. Figure 5 shows the isolines of $|\tau\cdot_{max}|$ representing the highest tangential stress of the three similar quantities τ_{12}, τ_{23} and τ_{31}. For the given plane strain problem it is calculated as follows:

$$|\tau\cdot_{max}| = (\sigma_2 - \sigma_1)/2 \text{ at } \sigma_1 > 0, \sigma_2 < 0, \quad (5)$$

$$|\tau\cdot_{max}| = \sigma_2/2 \text{ at } \sigma_1 < 0, \sigma_2 < 0. \quad (6)$$

We shall now examine two points in the vicinity of a fracture, one of which (A) is located in the region of lowered compressive stresses and the other (B), in the region of elevated compressive

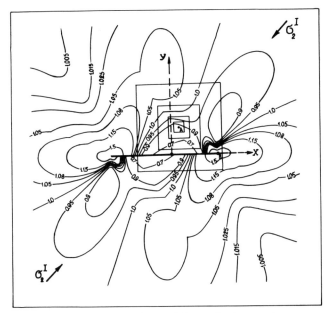

Fig. 5. Isolines of the modulus of the maximum tangential stresses $|\tau^*_{max}|$ for the local fracture field. A and B denote points for stress calculations at various integration distances. Quadrangles indicate projection of the edge of the volume studied on the drawing plane.

stresses. A mean stress field in progressively increasing volumes is calculated. Their projections on the plane of the drawing (squares) are presented in Figure 5 around the point A. The calculations are made by integration of the $|\tau^*_{max}|$ values in the unit volumes W whose centers are displaced in the system of the coordinates x and y:

$$|\overline{\tau^*_{max}}| = \frac{1}{N} \sum_{n=1}^{N} |\tau^*_{max}| (x,y) . \quad (7)$$

The value of z is taken to be equal to 1. Figure 6 shows the results of calculation of $|\overline{\tau^*_{max}}|$ for the points A and B as a function of a/L, where a is the side length of the square representing the volume of our study, projected on the plane of the drawing. It is evident from the figure that as a/L increases, one can observe a transition from the local lowered stress field of the II-nd rank to the external field of the I-st rank around the point A. Near the point B the transition is going from elevated local stress to the external field.

It is clear that under the real conditions, as characterized by the existence of active faults of various lengths and orientations in the Earth, some of the regions will have stress fields of varying ranks simultaneously.

The above example demonstrates a seemingly paradoxical situation. Depending on the measuring distances, one can obtain varying stresses for the same region of the space, and still all the results will be theoretically correct.

What is discussed above suggests that inhomogeneity of the stress-strain state of the rocks shows up at the different scales of the rock masses under study. One problem remaining to be solved is how continuous and to what order of magnitude will be the distribution of the homogeneous volumes in their sizes. Are there any discreteness and hierarchy in this distribution ?

The Scale of Block Sizes

Some insight into this problem can apparently be obtained from studies of the block structure of the rocks. A large body of the published data has been summarized by Sadovsky et al. [1982] to show the existence of characteristic, selected block sizes detected by use of a variety of geological, geodetic and geophysical techniques. This data, with some additions, are given in Table 1. The extreme left hand side column lists the length scale divided on subintervals differing by the factor of 10. The numbers in the right hand side part of the table describe the mean block sizes as revealed by:

1 - granulometric study of the samples from the bore pits; 2 - explosion crushing; 3 - spectral study by ultra-sounding; 4 - geophysical studies near dams; 5 - fluctuations of the seismic wave parameters in mines; 6 - studies of earthquake b-values; 7 - fluctuations of the seismic wave parameters; 8 - geological studies of the crustal

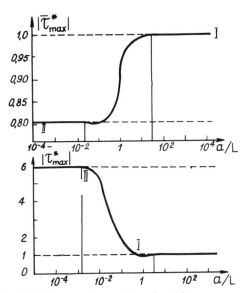

Fig. 6. Calculational results for the stress level $|\overline{\sigma^*_{max}}|$ at points A and B (top and bottom, respectively) at various a/L values. I and II are the fields of the I-st and II-nd rank, respectively.

TABLE 1. Sequence of Block Sizes in Rocks

Scale (m)	Results by various measurement methods (m)								
	1	2	3	4	5	6	7	8	9
10^{-3}=1mm	1.0								
	3.8	5.0							
10^{-2}=1cm	1.7	2.0							
		5.0							
		1.4							
10^{-1}		5.0	4.5	6.0					
10^{0}=1m		1.0	1.8	2.1					
					5.0				
10^{1}				0.9	1.75				
				5.0					
				2.0	2.0				
10^{2}						5.75			
10^{3} =1km							1.1		
10^{4}						4.25	5.25	6.1	4.0
							1.1	1.0	1.0
10^{5}						2.4	4.5		2.4
									1.15
10^{6}								4.0	4.4

block; 9 - studies of the distribution of the planets of the Solar system in size. All the values in the columns 1-9 are received from statistical analysis of distribution graphs of block sizes. Happened to be that in each subinterval differ by the factor of 10 (extreme left column) place not more than two block sizes. Of special interest are the values in columns 2, 4, 7, 9 showing the sequence of the 4 or 5 predominating block sizes by use of the same method. Comparison of the results obtained by the various independent methods seems to suggest that the ratio between each ensuing and preceding size $L = h_{i+1}/n_i \simeq 3.2$ can be derived for length-scales with 10 multiples.

In this context it should be noted that nearly the same ratio between the mean fracture-to-fracture distance and the mean fracture size in the strained bodies, known as "the concentration criterion" [Zhurkov et al., 1977; Sobolev and Zavialov, 1980], K = 3-5 takes place within a range of at least 10 orders of magnitude. Considering that the mean distance between fractures of the same rank controls the possible length of fractures of lower ranks, it may be assumed that there exists a profound relationship between the K values and the data of Table 1.

Thus an inhomogeneous hierarchical structure of the stress-strain fields is a remarkable property of rocks. Studies of the principles governing the distribution of these fields and their initiation represent important problems of the Earth's physics. The establishment of such a structure needs to use all the available methods permitting integration of the medium's properties of various sizes of the objects under study.

References

Markov, G. A., About the origin and regularities of horizontal stresses in the upper crust. Geotectonika, 3, 32-41, 1983.

Myachkin, V. I., D. N. Osokina, and N.Y. Zvetkova, Tectonophysical analysis of stress field and the problems of the physics of the source, The Models of Stress-strain State of Rock Massifs as related on Earthquake Prediction, Apatiti, pp. 3-24, 1982.

Osokina, D. N., and V. H. Fridman, The study of relation between fracture sides displacements and tectonic stresses of different levels, Recent Movements of the Earth Crust, Kishinev, pp. 89-91, 1982.

Ponomarev, V. S., Zoning relaxation of stresses during unloading of rock massifs, Report of Academy Science of USSR, 259, 1337-1339, 1981.

Riznichenko, Y. V., O. V. Soboleva, O.A. Kuchai et al., Seismotectonic strain of the Earth crust in the Middle Asia, Physics of the Earth, 10, 90-103, 1982.

Savich, A. I., V.I. Koptev, and Z. G. Yashchenko, The study of Elastic and Deformation Proerties of Rocks by Seismoacoustic Methods, Moscow, pp. 213, 1979.

Savich, A. I., and A. M. Zamachaev, The basic

regularities of natural stress distribution in the massif near Ingury dam, <u>Geological and geophysical investigation near Ingury dam</u>, Tbilisi, pp. 93-107, 1981.

Sadovsky, M. A., L. G. Bolchovitinov, and V. F. Pisarenko, About the inhomogeneity of rocks, <u>Physics of the Earth</u>, <u>12</u>, 2-18, 1982.

Sobolev, G. A., and A. D. Zavialov, The concentration criterion of seismoactive faults, <u>Report of Academy of Sciences of USSR</u>, <u>252</u>, 69-71, 1980.

Sobolev, G. A., A.A. Semerchan, B.G. Salov et al., The failure precursors of big rock sample, <u>Physics of the Earth</u>, <u>8</u>, 29-43, 1982.

Spetzler, H. A., G. A. Sobolev, C. H. Sondergeld et al., Surface deformation, crack formation and acoustic velocity changes in pyrophyllite under polyaxial loading, <u>J. Geophys. Res.</u>, <u>86</u>, No.B-2, 1070-1080, 1981.

Voevoda, O. D., The deformation of the Earth crust containing of faults, <u>Earthquake Precursors Modeling</u>, Moscow, 93-121, 1980.

Voevoda, O. D., About connection between residual desplacements, strain and tilts of the Earth surface with foci parameters, <u>Physics of the Earth</u>, <u>2</u>, 27-33, 1984.

Zurkov, S. N., V. S. Kuksenco, V.A. Petrov el al., About rock failure prediction, <u>Physics of the Earth</u>, <u>6</u>, 11-18, 1977.

COMPILATION OF EARTHQUAKE FAULT PLANE SOLUTIONS IN THE NORTH ATLANTIC AND ARCTIC OCEANS

Páll Einarsson

Science Institute, University of Iceland, 107 Reykjavík, Iceland

Abstract. A set of 95 fault plane solutions for earthquakes along the mid-oceanic ridge system in the North Atlantic and Arctic has been compiled, several of which are reported here for the first time. All the solutions are single event solutions, and most of them are based on teleseismic, long period data. Almost all fault plane solutions for earthquakes in fracture zones show transform faulting, sometimes with a component of reverse faulting. A majority of solutions for earthquakes in the axial zone of the ridges displays normal faulting. P-waves from these earthquakes usually have anomalous radiation pattern; the apparent nodal surfaces are not orthogonal planes. Interference between P, pP and sP is considered to offer the most likely explanation for this phenomenon. Solutions indicating reverse faulting are found near the ridge axis in two areas, near the Bárdarbunga Volcano in Central Iceland, and near 50°N. Volcanic processes such as the deflation of a localized magma chamber may cause this type of faulting, even in the extensional tectonic regime near a divergent plate boundary.

Introduction

Fault plane solutions of earthquakes play a major part in the theory of plate tectonics. Slip vectors of earthquakes along plate boundaries provide information on the direction of relative motion of the plates on either side, and focal mechanisms of intraplate earthquakes constrain possible models of the driving mechanism of the plates. The implementation of the World Wide Standardized Seismograph Network in the early sixties made available large quantity of high quality long period data with world wide coverage. It then became possible to determine fault plane solutions on a routine basis for most earthquakes larger than magnitude (m_b) 5 1/2, even in remote areas such as the mid-oceanic ridges in the Atlantic and arctic Oceans. Since then a multitude of papers have been published describing seismicity and fault plane solutions of major plate boundaries. The present paper grew out of an effort to summarize research on the seismicity along the boundary between the North American and Eurasian plates [Einarsson, 1985] shown in Figure 1. Collecting and editing previously published work on fault plane solutions along this boundary proved to be a nontrivial task. Furthermore, it turned out that substantial amount of data existed that had not been analyzed for fault plane solutions. This report therefore contains fault plane solutions for several earthquakes that have not been studied before, in addition to a compilation of previously reported solutions. It is hoped that the outcome is a relatively homogeneous data base, that can be used in future studies of plate motions and dynamics. The results are given in the Appendix, along with first motion plots of solutions presented in this paper. All solutions are shown schematically on seismicity maps in Figures 2-9.

Previous Studies

One of the most difficult parts of a compilation work is to select the material to be included. The papers on the subject differ enormously with regard to the methods used, amount of work and thoroughness of documentation, and it becomes inevitable to pass a judgement on the reliability of the published work. Following guidelines were used in the selection:

1. The fault plane solution is based primarily on long-period data, i.e. the wavelength is comparable to the source dimensions. Short period data are often difficult to use and lead to inconsistencies [Stefánsson, 1966; Sykes, 1967].

2. The seismograms should preferentially be read by the author. First motions reported in bulletins are often inconsistent, and turn out to be in error.

3. If the solution is based on P-wave first motion, a plot of the focal sphere should be included, so that the reader may judge the reliability of the solution.

4. Composite fault plane solutions, i.e. where data for more than one event are needed to obtain a solution, are rejected.

If more than one fault plane solution was avail-

Copyright 1987 by the American Geophysical Union.

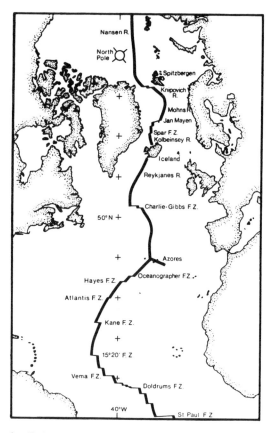

Fig. 1. Index map of the plate boundaries in the North Atlantic and Arctic Oceans.

able for the same event, the one was selected that most closely conformed to the guidelines. Strict use of the criteria excludes all earthquakes before the installation of the WWSSN. If not given in the original publications, poles of nodal planes, P- and T-axis, as well as O-axis of the solutions were derived and tabulated. Small and obvious errors were corrected in a few instances.

During the last few years focal mechanisms based on moment tensor inversion have been published in the Preliminary Determination of Epicenters listings of the U.S. Geological Survey, including several earthquakes in our study area. Awaiting further documentation these mechanisms have been omitted from the present compilation.

New Solutions

All new fault plane solutions reported here are based primarily on first motions of P-waves. In a few cases S-wave polarization was determined to better constrain the solution. All records were read by the author. First the arrival time of the P-wave was read from the short period record, and then the first motion was determined from the long period record at the appropriate time. By this procedure it is felt that the danger of erroneous picking of the first P-wave excursion is minimized. The rays were projected back to the focal sphere in a standard way using the Herrin tables. The earthquakes were assumed to originate in the lower part of the crust. Focal sphere plots of the new solutions are shown in the Appendix. Data from the WWSSN and Canadian network were used exclusively.

Non-Orthogonal Nodal Planes

Most authors studying focal mechanisms of mid-oceanic ridge earthquakes have come across the phenomenon that for some earthquakes the dilatational and compressional fields of the focal sphere cannot be separated by two orthogonal planes. This is observed for nearly all earthquakes in the axial zones of the Mid-Atlantic Ridge for which sufficient data are available. In the fracture zones, on the other hand, no examples have been found so far. Several explanations have been suggested for this phenomenon:

1. The apparent nonorthogonality is the result of the bending of rays passing through a low velocity lens beneath the ridge axis [Solomon and Julian, 1974].

2. The anomalous radiation pattern is the effect of near-source anisotropy related to preferred orientation of olivine crystals in the ridge mantle [Kawasaki and Tanimoto, 1981].

3. The earthquake source has an explosive component superimposed on the double couple [Solomon and Julian, 1974].

4. The earthquake occurs by extension failure of a porous, fluid-saturated medium [Robson et al, 1968]. The opening of an extension crack is accompanied by a sudden drop in fluid (magma) pressure. Thus an implosive component is superimposed on the compressive field from the extension crack.

5. The first motion pattern is obscured by the surface reflected phases pP and sP [Hart, 1978; Tréhu et al., 1981]. For a shallow source the arrival time difference between the reflected phases and the direct P-wave may be short compared to the response time of the long period seismographs. This effect would only occur for earthquakes with a significant dip-slip component.

All these suggestions deserve serious attention. The last explanation is favoured here, mainly because of the convincing arguments presented by Tréhu et al. [1981]. They inverted the Rayleigh waves from an event on the Reykjanes Ridge, for which non-orthogonal nodal planes had been found, and derived the moment tensor. The source mechanism turned out to be primarily of the double-couple type. They then calculated synthetic seismograms taking into account the surface reflected phases and the response of the seismographs, that showed remarkable similarity with actual data. The fourth possibility may also deserve special investigation in the light of recent findings of Julian [1983] for the Mammoth Lake events of 1980 associated with renewed activity of the Long Valley Caldera, and Foulger

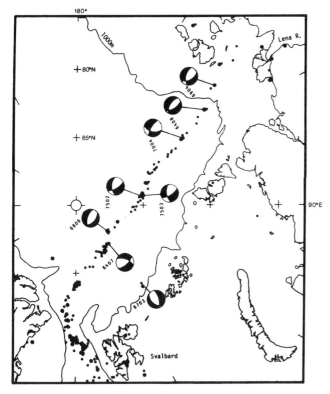

Fig. 2. Epicenters and fault plane solutions in the Arctic Basin. Epicenters are taken from the PDE lists of USGS for the period 1962-1981, only epicenters determined with 10 or more stations are included. Epicenters of events of m_b = 5.0 and larger are marked with large dots. The fault plane solutions are shown schematically as lower hemisphere stereographic projections of the focal sphere. The compressional quadrants (containing the least compressive stress axis) are shown black. Numbers give year and month of the earthquake.

[1984] for some earthquakes in the Hengill geothermal area in Iceland. A compensated linear vector dipole (CLVD) model seems to best conform with their data. This corresponds to an extensional crack with a superimposed pressure drop. The projections of the nodal surfaces onto the focal sphere are small circles centered on the T-axis.

Taking away the constraint of orthogonality of nodal planes it becomes impossible to obtain a focal mechanism solution for most earthquakes from first motions of P-waves alone. More sophisticated methods have to be applied. This constitutes a problem in this compilation work. Some of the fault plane solutions of normal faulting earthquakes reported in the literature are obtained with limited data, leaning heavily on the assumption of orthogonal nodal planes. Accepting minor inconsistencies, it is in many cases possible to find orthogonal planes if the station distribution on the focal sphere is uneven. This should be born in mind when reading the fault plane solution maps in Figures 2-9 and the table in the Appendix. In spite of this problem it is in many cases possible to infer the type of faulting and the general attitude of the T-axis from the first motion pattern, even if the exact position of the nodal surfaces cannot be determined.

The Arctic Plate Boundary

We will now trace the plate boundary from the continental shelf of Siberia near the Lena River delta, across the Arctic Basin, past Svalbard, along the Knipovich and Mohns Ridges, the Jan Mayen Fracture Zone, and the Kolbeinsey Ridge, to Iceland.

On the continental shelf of Siberia, the seismicity of the plate boundary is dispersed over a wide area, probably reflecting the continental lithospheric structure as well as the proximity to the pole of relative plate rotation. This area was noted for its relatively large earthquakes already by Tams [1927]. A fault plane solution by Conant [1972] shows normal faulting with non-orthogonal nodal planes. This is noteworthy because it shows that this phenomenon can occur in continental environment.

As the late boundary crosses the continental shelf edge over to the oceanic structure the seismic belt becomes narrow. It follows the crest of the Nansen (Gäkkel) Ridge for about 2000 km across the Arctic Basin (Figure 2). The ridge is almost straight, uninterrupted by large offset transform faults. The seismicity is moderately high and all available fault plane solutions show normal faulting at the ridge crest [Savostin and Karasik, 1981; Sykes, 1967].

North-east of Greenland the plate boundary makes a turn and follows the Lena Trough southwards to the Spitsbergen Fracture Zone (Figure 3). The Lena Trough appears to be an obliquely spreading boundary, similar to the Knipovich Ridge farther south. No bathymetric or seismological evidence for transform faulting along these boundaries has been found so far. In the Spitsbergen Fracture Zone, on the other hand, the alignment of epicenters, fault plane solutions, the occurrence of large earthquakes and bathymetric features are consistent with transform faulting. The short Molloy Ridge connects it to the Molloy Fracture Zone, which has been seismically quiet for the last two decades.

The Svalbard Archipelago stands out on most seismicity maps as an area of high seismicity. Most of the activity is concentrated in two zones in Heer Land and Nordaustlandet although scattered activity is found in other parts of the archipelago [Bungum et al., 1982]. Teleseismic maps, such as Figure 3, show intraplate activity on the continental shelf west of Svalbard, possibly connecting it to the Knipovich plate boundary. The concentrated earthquake zones in Svalbard are elongate in a WNW-ESE direction and a fault plane

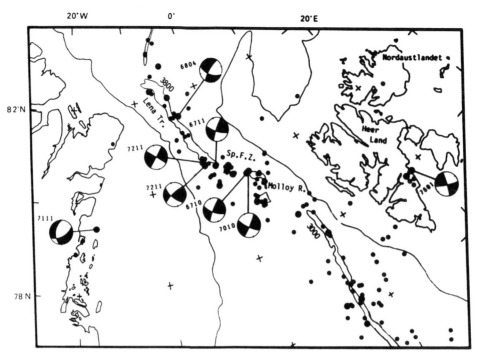

Fig. 3. Epicenters and fault plane solutions in the NE Greenland-Svalbard region. Data and symbols are as in Fig. 2. The continental shelf edge is marked by the 1000 m isobath, rift valleys and depressions associated with the plate boundary along the Nansen Ridge, Lena Trough, Spitsbergen and Molloy Fracture Zones and the Knipovich Ridge are shown with the 3800 and 3000 m isobaths.

solution shows left-lateral strike-slip along this trend. This seismicity was attributed by Bungum et al. [1982] to an interaction between the present tectonic stress field and older zones of weakness. Savostin and Karasik [1981], on the other hand, felt that this seismicity implied the existence of a separate plate, the Spitsbergen microplate, even though its eastern boundary could not be clearly delineated.

The seismicity of the Knipovich Ridge is quite scattered, implying that deformation takes place within a wide zone. This feature is possibly inherited from the time when this part of the plate boundary changed from being primarily of strike slip character to being a divergent boundary, when spreading ceased in the Labrador Sea.

The pattern of seismicity changes abruptly from the Knipovich to the Mohns Ridge. The Mohns Ridge is in most respects a typical mid-ocean ridge, centrally located in the Greenland-Norwegian basin, and uninterrupted by fracture zones of significant offset. The spreading is slightly oblique, especially at its western end, where it joins with the Jan Mayen Fracture Zone at an angle of 120 . The seismic zone is well defined, narrow and continuous along the rifted crest (Figure 4). Fault plane solutions by Conant [1972] and Savostin and Karasik [1981] show normal faulting at the ridge axis.

The Jan Mayen Fracture Zone is the most significant fracture zone in the Arctic region, displacing the ridge axis 210 km to the right. Both ridge-transform intersections are well identified in the topography by the characteristic triangular depressions. The transform section of the fracture zone is a pronounced trough, but highly asymmetric, because of the high topography connected with the Eggvin Bank and the Jan Mayen continental sliver to the south. The overall trend of the transform section is 110°, and all available fault plane solutions are consistent with left-lateral strike-slip along a plane with strike varying between 120° in the east to 102° in the west. This variation hardly justifies much speculation although some authors have felt that an explanation was needed. Bungum and Husebye [1977] suggested that the fracture zone consisted of several en echelon segments striking slightly more E-W than the main zone, and Savostin and Karasik [1981] concluded that the slip vector contained a small component of convergence across the fracture zone. Perhaps a stress concentration around an asperity can explain the variation. The asperity is the island of Jan Mayen that protrudes into the fracture zone about 55 km west of the eastern ridge intersection. The high seismicity of this part of the fracture zone may be an indication of stress concentration.

The plate boundary between Jan Mayen and Iceland follows the Kolbeinsey Ridge, which is

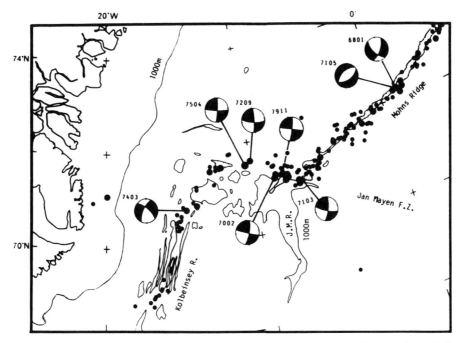

Fig. 4. Epicenters and fault plane solutions in the Jan Mayen area. Data and symbols are as in Fig. 2. The rift valley of the Mohns Ridge and depressions at the ridge-transform intersections are shown by the 2600 m isobath. The 1000 m isobath delineates the Jan Mayen Ridge, Eggvin Bank, and the Kolbeinsey Ridge.

anomalous in several ways. It is asymmetrically located with respect to the adjacent continents, the topography is high and the seismicity low. All these features may be in some way related to the existence of the Iceland hot spot to the south. Two small fracture zones are known on the Kolbeinsey Ridge, one located south of the Eggvin Bank near 71°N, the other near 69°N, the Spar Fracture Zone. These fracture zones are defined by topography and magnetic data, no fault plane solutions are available supporting transform faulting.

Iceland

Fault plane solutions are available for five areas, i.e. the two transform zones in South and North Iceland, the intraplate Borgarfjordur area in West Iceland, the Bárdarbunga Volcano in Central Iceland and the Katla Volcano in South Iceland (Figure 5). Solutions in the last two areas are reported here for the first time.

The plate boundary is displaced from the Kolbeinsey Ridge eastwards to the neovolcanic zone of Iceland by the Tjörnes Fracture Zone, which is a zone of high seismicity near the north coast of Iceland. The zone has a general E-W trend and fault plane solutions are consistent with transform faulting along the zone. In other respects the zone bears little resemblance to oceanic transform faults. The seismicity is distributed over an 80 km wide zone and is clearly not associated with slip along a single fault. At least three parallel, NW-SE trending seismic belts have been identified, which appear to take up the transform motion at present [Einarsson, 1976, 1979]. The three earthquakes for which fault plane solutions have been obtained did not occur on the same belt. The westernmost event (63 03) probably occurred on the southernmost belt, the Dalvik seismic line, the two other events were associated with the Grímsey seismsic line, which marks the northern boundary of the fracture zone. The easternmost event (76 01) originated at the intersection between the Grímsey line and the Krafla fissure swarm, and was associated with a rifting event along the divergent plate boundary [Björnsson et al., 1977]. It demonstrates the close association between plate separation along a ridge axis and transform faulting on the adjacent transform fault. The sequence of events involved a small basaltic eruption at the Krafla Volcano in the volcanic rift zone, deflation of the volcano, magma migration northwards along the Krafla fissure swarm, accompanied by rifting and large scale horizontal and vertical ground displacements. Transform faulting was initiated when the rifting reached the intersection with the Grimsey line. Activity has continued in the Krafla area since the initial events of December 1975 - January 1976, involving lateral migration of magma, eruptions and rifting.

A cluster of epicenters appears in Central Iceland in Figure 5. Most of this activity is as-

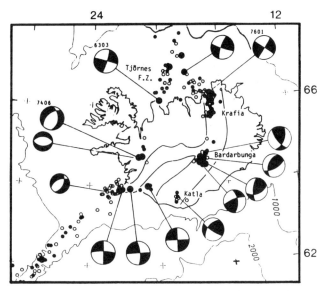

Fig. 5. Epicenters and fault plane solutions in Iceland. Data and symbols are as in Fig. 2.

sociated with the subglacial Bárdarbunga Volcano, which is one of the central volcanoes that characterize the structure of this part of the rift zone. A distinct sequence of earthquakes, with seven shocks reaching magnitude larger than 5 (m_b) occurred under Bardarbunga in 1974-1980, but events of that size were unknown there before. Four fault plane solutions have been determined, all showing reverse faulting. This result is somewhat unexpected, especially since Bardarbunga is located in a rift zone near the center of the Iceland hot spot [see e.g. Sigvaldason et al., 1974] where one would usually assume plate divergence and normal faulting. An inflating magma chamber would also cause extension and normal faulting in the crust above. Deflation, on the other hand, would lead to horizontal contraction and reverse faulting in the magma chamber roof. Seen in this perspective the increased seismic activity in 1974-1980 is an indication of decreasing pressure or magma withdrawal from underneath Bardarbunga. It is noteworthy in this context, that the Krafla Volcano in the rift zone about 110 km north of Bardarbunga was inflating during the same period. This may, of course, be coincidental, but it is also possible that some kind of a pressure connection between the volcanoes exists at subcrustal depths. The partially molten layer inferred from seismic and magneto-telluric data [Gebrande el al., 1980; Beblo and Björnsson, 1980] to lie beneath large parts of Iceland at the depth of 10-20 km could act as a conductor of pressure variations between volcanoes.

A group of epicenters is also seen in Figure 5 in the volcanic zone in South Iceland. These earthquakes are associated with the subglacial Katla Volcano, which is situated south of the ridge-transform intersection in South Iceland. One poorly constrained fault plane solution indicates strike-slip with a significant component of reverse faulting. As in the case of Bárdarbunga, a deflating magma chamber may offer an explanation for this type of faulting.

Most of the seismicity in SW-Iceland is attributed to a complex zone of transform faulting that connects the southern part of the volcanic zone to the submarine Reykjanes Ridge. The tectonic characteristics change along this zone, as one goes from the South Iceland seismic zone, that bridges the gap between the two branches of the volcanic zones, to the Reykjanes Peninsula, that displays high volcanic activity as well as high seismicity. Historically the largest earthquakes have occurred in the easternmost part of the zone; the maximum magnitude decreases as one goes westwards. Surface fracturing during historical earthquakes in the South Iceland seismic zone shown that each individual event is associated with right-lateral strike-slip faulting on N-S striking fault planes, in spite of the clear E-W alignment of the epicenters [Einarsson and Eiriksson, 1982; Einarsson et al., 1981]. The only available fault plane solution [Ward, 1971] in this part of the zone is consistent with this type of faulting. Most surface fractures on the Reykjanes Peninsula have a NE-SW trend, arranged en echelon with respect to the seismic zone, and appear to be related to magmatic activity. Two teleseismic fault plane solutions show strike-slip faulting on E-W or N-S striking planes [Ward, 1971; Einarsson, 1979]. Numerous solutions were also determined for small earthquakes by Klein et al., [1973, 1977] using dense, local seismic arrays. The T-axis of the solutions is consistently oriented in a horizontal, NW direction. The P-axis rotates between the vertical direction in normal faulting solutions and the horizontal NE direction in strike-slip solutions. Thus the stress regime is characterized by the NW-trending minimum compressive stress. The other principal stresses are probably nearly equal and may change directions according to local, or time dependent conditions. Dikes and eruptive fissures open up against the minimum stress and strike NE. There is a systematic trend along the peninsula from more strike-slip faulting in the eastern part to more normal faulting in the west, where the seismic zone gradually bends to the SW to join the Reykjanes Ridge.

A sequence of earthquakes occurred in West-Iceland in 1974, well outside the zones of rifting and volcanism normally attributed to the plate boundary. The sequence was studied by Einarsson et al. [1977], who determined a teleseismic fault plane solution for the largest shock, and several single event and composite solutions using data from a dense local array of seismographs. All solutions show normal faulting, indicating horizontal extension in this intraplate region.

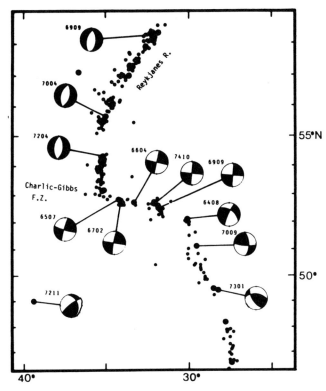

Fig. 6. Epicenters and fault plane solutions in the Charlie-Gibbs Fracture Zone and along the adjacent ridges. Data and symbols are as in Fig. 2.

From Iceland to the Azores

This section of the mid-Atlantic plate boundary consists of two gently arcuate ridges, offset near 52°N by a major fracture zone, the Charlie-Gibbs Fracture Zone. Nearest to Iceland the seismicity of the ridge is relatively low. Fault plane solutions have not been obtained for this reason. In this respect the Reykjanes Ridge shows a certain similarity with the Kolbeinsey Ridge north of Iceland. The low seismicity has been ascribed to the proximity to the Iceland hot spot [see e.g. Vogt, 1978]. Farther south the ridge crest gradually turns from a N35°E trend to a more southerly direction, and south of 58 1/2°N the seismicity attains levels comparable to other parts of the ridge. Four fault plane solutions are available all with non-orthogonal nodal planes. Two of them are for events in the same swarm (70 04), and are shown as one in Figure 6.

The Charlie-Gibbs Fracture Zone offsets the ridge crest about 350 km to the left. It consist of two parallel troughs, 45 km apart. A short spreading axis joins the troughs near 31°45'W, so transform faulting takes place in the western part of the northern trough and eastern part of the southern trough. This is confirmed by 5 fault plane solutions for the northern trough, but the southern trough has been seismically quiet for the past 25 years. One fault plane solution near the eastern ridge intersection shows oblique faulting. Oblique structures have been found near some ridge-transform intersections, e.g. the eastern Charlie-Gibbs [Searle, 1981], in the FAMOUS area [Whitmarsh and Laughton, 1975] and the Vema Fracture Zone [Forsyth and Rowlett, 1979].

The ridge segment south of the Charlie-Gibbs Fracture Zone, between 48° and 51°N, appears to have same peculiar features. A fault plane solution of an event near 49.5°N shows reverse faulting, and a second solution near 51 N has a significant component of reverse faulting. The seismic zone has a general trend of NNW, and according to Johnson and Vogt [1973] the structure of the ridge is characterized by alternating N-S trending and oblique spreading axes. The N-S axes are associated with transverse basement ridges, that trend slightly north of the spreading direction on both sides of the plate boundary. This "herringbone" pattern in the topography was interpreted as the result of asthenospheric flow southwards from the Iceland hot spot. The intersection of the transverse ridges with the plate boundary are then viewed as the locus of unusually high production of volcanic material, a central volcano complex, that slowly migrates southwards along the plate boundary, leaving a trail of its products on both plates. Thus the reverse faulting solutions may be put into a volcanic context as was the case with the Bárdarbunga and Katla earthquakes in Iceland. Forcible intrusion of viscous magma at shallow depth can cause thrust faulting in the adjacent region, and the deflation of a magma chamber will cause reverse faulting in the chamber roof.

South of 48°N the ridge is fairly straight and fracture zone offsets are too small to be resolved on the seismicity map (Figure 7). Seismic activity is relatively uniform, both in space and time, and earthquakes larger than $m_b = 5$ are rare. Two poorly constrained fault plane solutions indicate normal faulting.

The Eurasian-African plate boundary is marked by a seismic zone that extends from a triple junction with the Mid-Atlantic Ridge near the Azores to Gibraltar. The characteristics of the activity change as one goes along the zone. It can be divided into three segemnts according to its seismic characteristics. The westernmost segment extends from the junction, through the Azores and to the eastern end of the archipelago. The zone is relatively narrow, comparable to the zone on the Mid-Atlantic Ridge, and follows the ESE trend of the archipelago. Fault plane solutions show that strike-slip is the principal mode of faulting, but the nodal planes only rarely strike parallel to the trend of the zone. The largest earthquake of recent years in this area occurred on Jan. 1, 1980. The fault plane solution is of the strike-slip type (see Appendix), and the aftershock study of Hirn et al. [1980] revealed the NNW striking plane as the fault

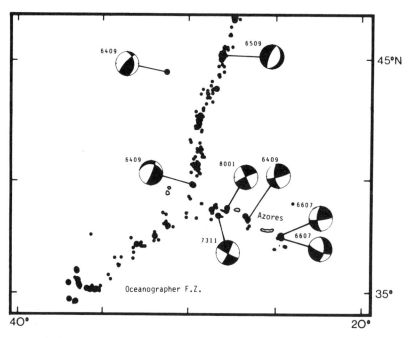

Fig. 7. Epicenters and fault plane solutions near the Azores triple junction. Data and symbols are as in Fig. 2.

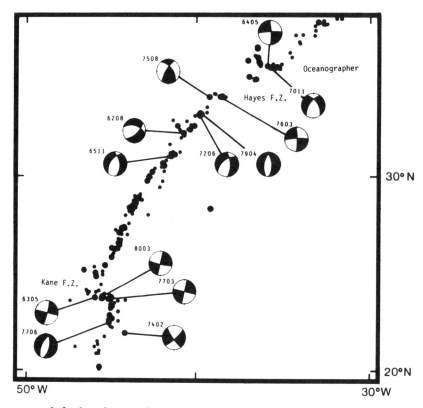

Fig. 8. Epicenters and fault plane solutions along the Mid-Atlantic Ridge between 20°N and 37°N. Data and symbols are as in Fig. 2.

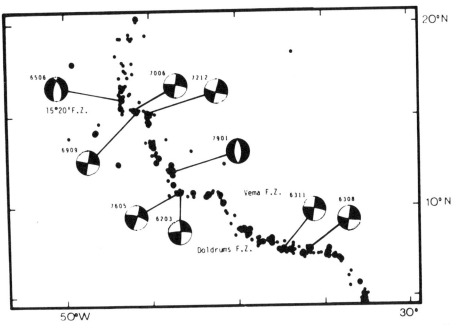

Fig. 9. Epicenters and fault plane solutions along the Mid-Atlantic Ridge between 5°N and 20°N. Data and symbols are as in Fig. 2.

plane, thus indicating left-lateral movement. If this result is generalized for the Azores seismic belt, it appears as if the relative plate motion is taken up by a series of en echelon strike-slip faults.

The middle section of the Azores-Gibraltar zone has been seismically quiet for the last decades. East of 18°W the plate boundary is no longer defined by a narrow seismic belt. The high, but diffuse seismicity in this region shows that plate deformation occurs within a 400-500 km wide zone, which extends into the Gulf of Cadiz and is connected to the seismic belt of Morocco and Algiers. Fault plane solutions are characterized by thrust faulting; occasionally strike-slip solutions are seen. A common feature is a maximum compressional axis trending N to NW, reflecting convergence between the Eurasian and African Plates.

The North American-African Plate Boundary

This part of the Mid-Atlantic Ridge is cut by unusually many large offset fracture zones. North of 25°N all the fracture zones offset the ridge crest to the right, but south of 25°N almost all offsets are to the left. This shapes the ridge system into an arcuate structure, concave to the east, reflecting the original shape of the continents at the time of break-up. Major transform faults can be identified on the seismicity maps (Figures 8 and 9) by one or more of their seismic characteristics, i.e. east-west alignment of epicenters, offset in the ridge crest seismic zone, and fault plane solutions indicating strike-slip faulting in the transform sense. Thus left-lateral slip is demonstrated in the zones where the ridge axis is offset to the right, such as in the Oceanographer Fracture Zone (1 solution) and the Hayes Fracture Zone (1 solution). Right-lateral faulting, on the other hand, is shown in the Kane (3 solutions), 15° 20'(3 solutions), Vema (2 solutions) and Doldrums Fracture Zones (2 solutions). The last one is a multiple fracture zone, consisting of 3-4 transform faults.

All fault plane solutions obtained for earthquakes at or near the ridge crest show normal faulting, and where sufficient data are available the solutions have nonorthogonal nodal planes. For this reason the orientation of the fault planes and the stress axes cannot be determined with confidence. Yet it is clear that these are variable, even for closely spaced events. The two normal faulting events near 33°N (Figure 8), for example, have nearly identical epicenters, but the first motion pattern of the P waves is considerably different.

Only two fault plane solutions deviate from the general pattern established above; one in the Oceanographer Fracture Zone, the other in the Hayes Fracture Zone. The former (70 11) appears to have a large component of normal faulting. Reexamination of the seismograms showed that this earthquake is a complicated event, with the first motion of the main event obscured by a small foreshock on many records. This can hardly be considered a reliable solution. The other solution shows a significant component of reverse faulting in the Hayes Fracture Zone. Large scale vertical movements are indicated by the presence of transverse ridges in some of the major fracture

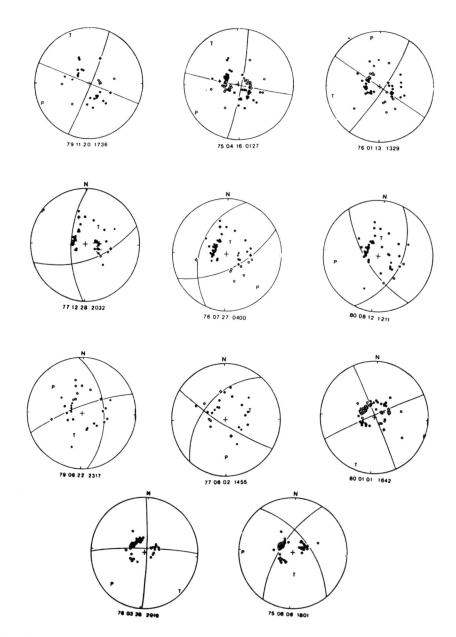

Fig. 10. Fault plane solutions shown on a stereographic projection of the lower hemisphere of the radiation field. Open circles are dilatational first motions of P-waves; compressional first motions are dots. Crosses indicate that the P-wave has a nodal character. S-wave polarization is shown with line segments. P and T are inferred axes of compression and tension, in the center of the dilatational and compressive fields of the focal sphere, respectively. Small symbols denote unreliable readings.

zones [Bonatti, 1978; Bonatti and Chermak, 1981; Bonatti et al., 1983] and seem to be an integral part of the tectonic regime of a transform fault. Occasional occurrence of reverse faulting events in fracture zones should therefore not be too surprising.

Conclusions

The area studied here contains the part of the mid-oceanic ridge system that is most easily studied by seismological methods because of its favorable location with respect to the dense seismic

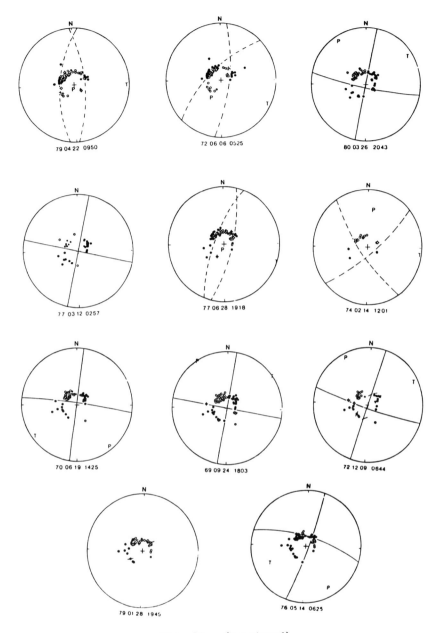

Fig. 10. (continued)

networks of Europe and N-America. It is here that we can hope to get the most complete picture of the faulting processes at mid-oceanic ridges by the study of earthquake focal mechanisms. The present compilation of fault plane solutions confirms the main results of Sykes [1967] that earthquakes in fracture zones are associated with transform faulting and that earthquakes along ridge axes are mainly associated with extensional tectonics. Thus nearly all fracture zone events, for which fault plane solutions have been obtained (35 of 37), were associated with transform faulting. For 25 out of 32 ridge crest events the fault plane solution had a large component of normal faulting. All the normal faulting events with sufficient focal sphere coverage have anomalous radiation pattern, i.e. orthogonal nodal planes cannot be found. Several explanations have been suggested for this phenomenon, but the one considered most likely is the interference of the surface reflected phases pP and sP with the direct P-wave. Fault plane solutions with large compo-

TABLE 1. Compilation of Fault Plane Solutions of North Atlantic and Arctic Earthquakes

Date Y M D	Time HM	Epicenter N	W	m_b	Poles of Nodal Planes Tr/Pl	Tr/Pl	P-axis Tr/Pl	T-axis Tr/Pl	O-axis Tr/Pl	Data Source
1969 04 07	2026	76.5	-130.8	5.5	234/28	102/32(1)	171/54	77/2	346/36	7(6,18)
1964 08 25	1347	78.1	-126.6	6.1	248/36	94/32(1)	165/71	262/2	351/19	24
1970 04 23	0055	80.7	-122.0	5.1	136/40	242/18	196/43	93/14	350/44	18
1975 02 26	0448	85.1	-98.0	5.4	182/42	64/28	112/52	214/10	312/36	18
1975 03 02	1423	85.0	-98.2	5.0	178/34	64/30	118/50	212/4	305/40	18
1968 06 08	0041	87.1	-51.3	5.2	158/36	6/50	109/75	349/6	258/13	18
1964 07 31	2345	86.3	-40.5	5.2	241/37	359/31	307/53	209/6	116/38	18
1967 03 14	0750	82.5	-40.5	4.6	82/55	297/31	344/68	101/12	195/16	18
1968 04 07	0516	81.5	3.9	5.2	36/46	132/7	94/40	352/16	245/46	18
1972 11 19	2010	80.4	2.6	5.3	44/4	134/2	359/1	89/5	260/85	18
1972 11 25	2003	80.3	2.4	5.5	56/30	154/14	11/12	108/32	264/55	18
1967 11 23	1342	80.2	1.0	5.7	38/10	130/6	352/6	85/16	242/72	12
1967 10 18	0111	79.8	-2.4	5.6	41/19	133/4	355/10	88/17	236/70	12
1970 10 26	2053	79.8	-2.5	5.5	48/14	317/4	1/14	93/6	213/76	7(18)
1971 11 26	2307	79.5	17.8	5.1	333/38	100/38	33/60	126/1	217/30	26
1976 01 18	0446	77.9	-18.6	5.6	30/20	298/8	75/2	343/15	171/76	5(15,18)
1970 10 21	0814	74.6	-8.4	5.5	43/29	300/26	349/39	82/3	175/51	18
1968 01 03	0737	72.2	-1.2	5.3	300/30	52/32	354/46	86/1	178/44	18
1971 05 31	0346	72.2	-1.2	5.5	322/26	132/36(1)	12/80	138/4	229/7	7(18)
1975 01 20	1047	71.8	-14.6	5.1	106/28	199/7	154/35	59/14	300/61	18
1971 03 23	0926	71.0	7.0	6.0	30/16	296/18	253/1	344/24	158/67	7(18)
1979 11 20	1736	71.2	8.0	5.6	22/0	292/5	247/3	337/3	112/85	2
1970 02 22	2341	71.1	8.6	5.2	21/3	290/16	244/9	336/14	119/73	18
1972 09 08	1134	71.6	10.0	5.7	12/20	280/10	56/4	322/20	163/67	18
1975 04 16	0127	71.5	10.4	6.1	12/3	282/8	237/3	327/8	121/82	2(5,18)
1974 03 22	1910	70.7	14.7	5.0	222/24	116/32	170/40	78/4	342/50	18
1969 05 05	2147	66.8	18.2	5.1	22/8	115/18	159/8	67/19	270/70	7
1963 03 28	0015	66.3	19.8	5.5	197/3	287/12	333/7	241/10	88/77	24(23,22)
1976 01 13	1329	66.2	16.7	6.0	212/0	302/10	348/7	257/7	122/80	2(5,18)
1974 06 12	1755	64.8	21.2	5.2	132/34	268/48	187/64	313/11	28/24	9
1974 07		64.8	21.4	-	0/45	180/45	-/90	0/0	90/0	9,3
1977 12 28	2032	64.7	17.3	5.0	341/32	92/30	306/1	38/47	215/44	2
1976 07 27	0400	64.7	17.4	5.2	333/30	110/52	134/13	17/64	230/20	2
1980 08 12	1211	64.7	17.3	5.2	291/47	50/25	255/12	3/55	158/32	2
1979 06 22	2317	64.6	17.4	5.4	159/15	265/45	309/18	199/45	54/41	2
1967 07 27	0517	64.0	20.7	5.0	0/0	90/0	45/0	135/0	-/90	29
1968 12 05	0944	63.9	21.7	5.5	357/0	267/15	221/9	313/10	87/75	29(18)
1973 09 15	0145	63.9	22.2	5.3	180/12	271/2	226/10	134/8	9/78	8
1972 09		63.8	22.7	-	153/42	296/42	225/70	135/0	45/20	14
1977 06 02	1455	63.7	19.1	4.9	32/10	132/45	180/23	70/38	293/44	2
1969 09 20	0508	58.3	32.2	5.5	(1)		172/72	278/5	9/15	8(20,11)
1970 04 24	0123	55.7	35.0	5.3	(1)		-/90	293/0	13/0	27(8)
1972 04 03	1852	54.3	35.1	5.3	(1)		-/90	95/0	5/0	27(8)
1972 04 03	2036	54.3	35.1	5.1	(1)		-/90	95/0	5/0	27(8)
1965 07 05	0831	52.9	34.2	5.7	16/10	107/8	331/1	61/12	235/77	25
1967 02 13	2414	52.8	34.1	5.5	10/0	100/6	146/4	56/4	280/84	8(19,13)
1966 04 08	0552	52.7	33.3	5.2	(12/0)	(102/0)	(147/0)	(57/0)	(-/90)	8
1974 10 16	0545	52.6	32.1	5.8	5/2	95/0	320/1	50/1	185/88	13
1969 09 24	0358	52.5	31.9	5.1	4/0	94/0	139/0	49/0	-/90	8
1964 08 26	0318	52.1	30.1	5.3	214/39	116/9	157/33	261/20	16/50	8
1970 09 18	1612	51.1	29.6	5.1	6/10	270/32	224/16	323/30	108/56	8
1973 01 05	0144	49.5	28.2	5.3	55/32	183/45	212/6	109/61	305/29	8
1972 11 07	1205	49.0	39.4	5.1	253/42	142/22	294/15	184/50	35/37	28
1965 09 29	2320	45.2	28.2	5.3	113/16	(1)	(113/68)	(293/22)	(23/0)	8(28)
1964 09 17	1502	44.5	31.3	5.5	263/59	132/22	296/21	166/20	34/21	26(28)
1964 09 18	1312	39.8	29.7	5.5	292/6	188/64	268/45	135/35	25/25	28
1973 11 23	1336	38.5	28.3	5.0	27/4	297/4	252/5	342/5	162/85	17
1980 01 01	1642	38.8	27.8	6.0	243/2	153/4	107/1	199/4	353/86	2(17)
1964 09 06	1855	38.3	26.6	4.8	81/30	346/10	126/14	28/29	240/57	28
1966 07 04	1215	37.5	24.7	5.3	22/20	276/36	334/40	235/10	132/48	4(16)
1966 07 05	0509	37.6	24.7	5.0	89/30	354/10	135/13	37/29	248/58	28
1975 05 26	0911	36.0	17.6	6.7	120/22	225/25	171/38	263/2	355/54	17
1975 03 08	0840	38.6	14.8	4.7	185/12	311/75	353/33	198/54	93/12	17
1970 12 30	2057	37.2	14.3	5.0	158/10	278/74	326/36	174/52	65/14	17
1969 07 26	1224	43.7	14.6	4.7	159/8	295/76	331/34	170/54	67/10	17
1969 09 06	1430	36.9	11.9	5.7	209/2	120/10	165/10	73/4	313/80	17(16,28)

TABLE 1. (continued)

Date Y M D	Time HM	Epicenter N	W	m_b	Poles of Nodal Planes Tr/Pl	Tr/Pl	P-axis Tr/Pl	T-axis Tr/Pl	O-axis Tr/Pl	Data Source
1972 04 18	0551	36.4	11.1	4.7	196/12	292/22	335/6	242/25	77/64	17
1962 12 26	0858	39.3	10.6	5.0	344/19	85/24	126/3	33/32	240/58	17
1969 02 28	0240	36.0	10.6	7.3	142/40	0/46	340/4	78/72	249/18	17(16,10,28)
1969 02 28	0425	36.2	10.5	5.7	142/30	0/54	337/12	100/68	243/18	17(16)
1964 03 15	2230	36.2	7.6	6.2	335/10	206/74	165/34	320/54	66/12	28(4,16)
1964 05 17	1926	35.2	35.9	5.5	176/6	267/16	219/16	311/7	64/73	24(28,30)
1970 11 18	1223	35.1	35.7	5.3	233/30	123/30	181/45	88/1	358/45	28
1976 03 28	2019	33.8	38.6	5.5	177/8	268/5	224/10	131/3	30/80	2
1975 08 06	1801	33.8	39.3	5.4	229/34	120/26	267/4	173/46	359/45	2
1979 04 22	0950	33.0	39.7	5.7	(1)		195/80	92/3	2/10	2
1972 06 06	0525	33.0	39.9	5.5	(1)		212/51	117/3	24/40	2
1962 08 06	0135	32.0	40.8	-	196/40	319/33	263/57	166/4	73/33	24(21)
1965 11 16	1524	31.0	41.5	6.0	(1)		193/60	283/0	13/30	24(30)
1963 05 19	2135	23.8	46.0	5.8	13/0	284/3	329/2	239/2	103/87	24(22)
1980 03 26	2043	23.9	45.6	5.9	11/8	101/0	325/6	56/6	191/82	2
1977 03 12	0257	23.8	45.2	5.4	(13/0)	(103/0)	(328/0)	(58/0)	(-/90)	2
1977 06 28	1918	22.6	45.1	5.8	(1)		202/79	108/1	18/11	2
1974 02 14	1201	22.0	44.3	5.4	54/18	320/15	7/24	97/1	191/68	2
1965 06 02	2340	15.9	46.6	5.7	(1)		158/64	269/9	3/24	25(30)
1970 06 19	1425	15.4	46.0	5.5	188/5	98/0	143/4	233/4	8/85	2(30)
1969 09 24	1803	15.2	45.8	5.7	10/0	100/0	315/0	55/0	-/90	2
1972 12 09	0644	15.2	45.2	5.6	16/7	106/0	330/4	61/4	196/83	2
1979 01 28	1945	11.9	43.7	5.8			(-/90)	(90/0)	(0/0)	2
1976 05 14	0625	10.8	43.5	5.6	195/22	287/6	149/10	243/20	33/67	24(23)
1962 03 17	2047	10.9	43.2	-	180/4	270/6	315/0	225/7	53/83	24(22)
1963 11 17	0048	7.6	37.4	5.8	188/4	98/0	143/2	233/2	8/86	24(22)
1963 08 03	1021	7.7	35.8	6.0	10/11	100/0	325/8	55/8	190/79	25
1965 08 16	1236	-0.5	20.0	6.1	354/16	85/7	316/5	41/18	199/72	24
1965 11 15	1118	-0.2	18.7	5.6	357/20	88/6	311/9	44/18	197/70	24

nents of reverse faulting were found at the ridge axis in two areas; in the Bardarbunga central volcano in Iceland and near 50°N. Several mechanisms may be found to explain reverse faulting at a divergent plate boundary, but the one favoured here is brittle failure of the crust above a deflating, localized magma chamber.

Appendix 1. New and Revised Fault Plane Solutions

Focal sphere plots of P-wave first motions and S-wave polarizations are shown in Fig. 10. Nodal planes, T- and P-axes are shown when they could be determined with reasonable accuracy. For normal faulting earthquakes the apparent nodal planes are probably not planar, although they have been approximated by planes.

79 11 20 1736h, Jan Mayen Fracture Zone. This is a reasonably well determined strike-slip solution in spite of relatively high microseismic disturbance at N-American stations. The sense of motion is left-lateral if the slip is along the fracture zone.

75 04 16 0127h, Jan Mayen Fracture Zone. This is a well constrained strike-slip solution, left-lateral if slip occurs along the fracture zone.

76 01 13 1329h, Tjornes Fracture Zone. This is a well constrained strike-slip solution, right-lateral if slip occurs along the Grímsey seismic lineament.

77 12 28 2032h, Bárdarbunga Central Volcano. This is clearly a reverse faulting solution. One nodal plane (strike 71°) is well determined, the other s determined mostly by the constraint of orthogonality. Assuming shallower focus would allow more freedom in strike, e.g. T-axis could be closer to vertical, which fits better to the S-wave observations.

76 07 27 0400h, Bárdarbunga Central Volcano. This clearly is a reverse faulting solution. One nodal plane (strike 64°) is well constrained, the dip of the other one mainly found by the constraint of orthogonality, its strike is affected by the assumed depth of focus and the local velocity model.

80 08 12 1211h, Bárdarbunga Central Volcano. The solution has a large component of reverse faulting. One nodal plane (strike 21°) is well determined, the other is found by the constraint of orthogonality.

79 06 22 2317h, Bárdarbunga Central Volcano. The solution has a significant reverse faulting component, but the attitude of the nodal planes is highly dependent on the assumption of orthogonality.

77 06 02 1455h, Katla Central Volcano. The solution indicates a mixture of strike-slip and

reverse faulting. No solution free of inconsistencies could be found. The preferred solution is a compromise.

80 01 01 1642h, Azores. A pure strike-slip solution fits the first motion data best. The sense of motion is left-lateral, if slip is along the plane striking 153° as shown by aftershocks.

76 03 28 2019h, Hayes Fracture Zone. The solution shows strike-slip faulting, left-lateral motion if slip along the fracture zone is assumed.

75 08 08 1801h, Hayes Fracture Zone. This solution has a significant component of reverse faulting. One nodal plane (strike 30°) is reasonably well constrained by the N.-American stations. The other plane is determined by one nodal reading and the constraint of orthogonality.

79 04 22 0950h, Ridge axis near 33° N. This is a normal faulting solution with nonorthogonal nodal surfaces. Many of the dilatational first motions have a nodal character.

72 06 06 0525h, Ridge axis near 33° N. This is a normal faulting solution with nonorthogonal nodal surfaces. In spite of nearly identical epicenter to the previous event, the first motion pattern of the two earthquakes is different.

80 03 26 2043h, Kane Fracture Zone. This is a well controlled strike-slip solution, right-lateral motion if slip occurs along the fracture zone.

77 03 12 0257h, Kane Fracture Zone. A strike-slip solution with right-lateral slip along the fracture zone fits the data well. The solution is highly nonunique, however. The strike of the nodal planes can be varied as much as 10°, and a thrust mechanism is possible with only minor inconsistencies.

77 06 28 1918h, Ridge axis near 22.6° N. This event is part of a large earthquake swarm. Three events were analyzed; at 1538, 1618 and 1918h. Judged from the long period body and surface waves, all the events had similar faulting mechanism. Small differences were seen in the short period wave forms. The best solution was obtained for the 1918 event. It is a normal faulting solution with non-orthogonal nodal surfaces. Most dilatational first motions have a nodal character.

74 02 14 1201h, Intraplate event. This solution is not well constrained, but is probably of the strike-slip type. The earthquake appears to be a typical intraplate event with small surface waves, and impulsive short-period P-waves. Most of the first motions are read from short-period records.

70 06 19 1425h, 15° 20' Fracture Zone. This is a strike-slip solution, right-lateral motion if slip is along the fracture zone. Because of the concentration of stations near the center of the plot, dips of the nodal planes are well determined, but the strikes are not. A nodal plane striking parallel to the fracture zone fits the data well, but the strike could be varied as much as 10°.

69 09 24 1803h, 15° 20' Fracture Zone. This is a well controlled strike-slip solution, right-lateral motion if slip is along the fracture zone.

72 12 09 1644h, 15° 20' Fracture Zone. This is a strike-slip solution similar to the two previous ones. Nodal plane of the preferred solution has a strike of 106°, but is not well constrained. The strike can be varied within a 23° angle without adding inconsistencies.

79 01 28 1945h, Ridge axis near 12° N. This event is the largest in a swarm that lasted at least 14 hours. The wave form and pattern of first motions indicate normal faulting, but the attitude of the nodal surfaces cannot be determined. The S-wave polarizations indicate a nearly vertical P-axis.

76 05 14 0625h, Vema Fracture Zone. This is a strike-slip solution with a small component of reverse faulting. The sense of motion is consistent with transform faulting even though the preferred nodal plane strikes slightly oblique to the fracture zone and dips to the north. This event is located near the ridge - fracture zone intersection.

Appendix 2. Compilation of Fault Plane Solutions

Earthquakes for which fault plane solutions have been determined are listed in geographical order in Table 1, along with parameters of the solutions. Reference to the data source is given in the last column by a number, alternate solutions or reference to further studies are given in parentheses as follows:

1. Nonorthogonal nodal planes
2. This paper
3. Average solution for a number of small earthquakes, determined with a dense, local array
4. Banghar and Sykes, 1969
5. Bungum, 1978
6. Chapman and Solomon, 1976
7. Conant, 1972
8. Einarsson, 1979
9. Einarsson et al., 1977
10. Fukao, 1973
11. Hart, 1978
12. Horsfield and Maton, 1970
13. Kanamori and Stewart, 1976
14. Klein et al., 1977
15. Mitchell et al., 1977
16. McKenzie, 1972
17. Moreira, 1982
18. Savostin and Karasik, 1981
19. Solomon, 1973
20. Solomon and Julian, 1974
21. Stauder and Bollinger, 1964
22. Stauder and Bollinger, 1966
23. Stefánsson, 1966
24. Sykes, 1967
25. Sykes, 1970
26. Sykes and Sbar, 1974
27. Tréhu et al., 1981
28. Udias et al., 1976
29. Ward, 1971
30. Weidner and Aki, 1973

Note added in proof. Since the acceptance of this paper for publication, significant progress has been made in the study of source mechanisms of mid-oceanic earthquakes. A few remarks in the paper may therefore seem out of place.

Acknowledgements. Financial contribution was obtained from the Icelandic Science Fund for this work. The WWSSN film chips library at Lamont-Doherty Geological Observatory was used extensively; the help and hospitality of Drs. R. Bilham and D. Simpson is gratefully acknowledged. Sigfús Johnsen, Sigurdur E. Pálsson and Bryndís Brandsdóttir helped with computing and plotting, Kristín Pálsdóttir typed the manuscript.

References

Banghar, A.R., and L.R. Sykes, Focal mechanisms of earthquakes in the Indian Ocean and adjacent regions, J. Geophys. Res., 74, 632-649, 1969.

Beblo, M., and A. Björnsson, A model of electrical resistivity beneath NE-Iceland, correlation with temperature, J. Geophys., 47, 184-190, 1980.

Björnsson, A., K. Saemundsson, P. Einarsson, E. Tryggvason, and K. Gronvold, Current rifting episode in north Iceland, Nature, 266, 318-323, 1977.

Bonatti, E., Vertical tectonism in oceanic fracture zones, Earth Planet. Sci. Letters, 37, 369-379, 1978.

Bonatti, E., and A. Chermak, Formerly emerging crustal blocks in the equatorial Atlantic, Tectonophysics, 72, 165-180, 1981.

Bonatti, E., R. Sartori, and A. Boersma, Vertical crustal movements at the Vema Fracture Zone in the Atlantic: Evidence from dredged limestones, Tectonophysics, 91, 213-232, 1983.

Bungum, H., Reanalyzation of three focal mechanism solutions for earthquakes from Jan Mayen, Iceland and Svalbard, Tectonophysics, 51, T15-T16, 1978.

Bungum, H., B.J. Mitchell, and Y. Kristoffersen, Concentrated earthquake zones in Svalbard, Tectonophysics, 82, 175-188, 1982.

Bungum, H., and E.S. Husebye, Seismicity of the Norwegian Sea: The Jan Mayen Fracture Zone, Tectonophysics, 40, 351-360, 1977.

Chapman, M.E., and S.C. Solomon, North American-Eurasian plate boundary in Northeast Asia, J. Geophys. Res., 81, 921-930, 1976.

Conant, D.A., Six new focal mechanism solutions for the Arctic and a center of rotation for plate movements, M.A. Thesis, Columbia University, New York, 1972.

Gebrande, H., H. Miller, and P. Einarsson, Seismic structure of Iceland along RRISP-Profile I, J. Geophys., 47, 239-249, 1980.

Einarsson, P. Relative location of earthquakes within the Tjörnes Fracture Zone, Soc. Sci. Isl., Greinar V, 45-60, 1976.

Einarsson, P. Seismicity and earthquake focal mechanisms along the mid-Atlantic plate boundary between Iceland and the Azores, Tectonophysics, 55, 127-153, 1979.

Einarsson, P., Seismicity along the eastern margin of the North American Plate. Manuscript, to be published by the Geol. Soc. America, 1985.

Einarsson, P., and J. Eiríksson, Earthquake fractures in the districts Land and Rangárvellir in the South Iceland Seismic Zone, Jökull, 32, 113-120, 1982.

Einarsson, P., S. Björnsson, G. Foulger, R. Stefánsson, and p. Skaftadottir, Seismicity pattern in the South Iceland seismic zone, Earthquake Prediction - An International Review, Maurice Ewing Series 4, Am. Geophys. Union, p. 141-151, 1981.

Einarsson, P., F.W. Klein, and S. Björnsson, The Borgarfjördur earthquakes of 1974 in West Iceland, Bull. Seism. Soc. Am., 67, 187-208, 1977.

Forsyth, D., and H. Rowlett, Microearthquakes and recent faulting at the intersection of the Vema fracture zone and the Mid-Atlantic Ridge (abstract), EOS Trans. AGU, 60, 376, 1979.

Foulger, G.R., The Hengill geothermal area: Seismological studies 1978-1984, Science Institute, Univ. Iceland, Rep. RH-07-84, 96 pp., 1984.

Fukao, Y., Thrust faulting at a lithospheric plate boundary, the Portugal earthquake of 1969, Earth Planet. Sci. Lett., 18, 205-216, 1973.

Hart, R.S., Body wave observations from the September, 1969, North Atlantic Ridge earthquake (abstract), EOS, Trans. Amer. Geophys. Union, 59, 326, 1978.

Hirn, A., H. Haessler, P. Hoang Trong, G. Wittlinger, and L.A. Mendes Victor, Aftershock sequence of the January 1st, 1980, earthquake and present-day tectonics in the Azores, Geophys. Res. Lett., 7, 501-504, 1980.

Horsfield, W.T., and P.I. Maton, Transform faulting along the De Geer line, Nature, 226, 256-257, 1970.

Johnson, G.L., and P.R. Vogt, Mid-Atlantic ridge from 47° to 51° N, Geol. Soc. Amer. Bull., 84, 3443-3462, 1973.

Julian, B.R., Evidence for dyke intrusion earthquake mechanisms near Long Valley Caldera, California, Nature, 303, 323-325, 1983.

Kanamori, H., and G.S. Stewart, Mode of strain release along the Gibbs fracture zone, mid-Atlantic ridge, Phys. Earth Planet. Interiors, 11, 312-331, 1976.

Kawasaki, I., and T. Tanimoto, Radiation patterns of body waves due to the seismic dislocation occurring in an anisotropic source medium, Bull. Seismol. Soc. Amer., 71, 37-50, 1981.

Klein, F.W., P. Einarsson, and M. Wyss, Microearthquakes on the mid-Atlantic plate boundary on the Reykjanes Peninsula in Iceland, J. Geophys. Res., 78, 5084-5099, 1973.

Klein, F.W., P. Einarsson, and M. Wyss, The Reykjanes Peninsula, Iceland, earthquake swarm of September 1972 and its tectonic significance, J. Geophys. Res., 82, 865-888, 1977.

McKenzie, D., Active tectonics of the Mediterranean region, Geophys. J. R. Astr. Soc., 30, 109-185, 1972.

Mitchell, B.J., J.E. Zollweg, J.J. Kohsmann, C.-C. Cheng, and E.J. Hang, Intraplate earthquakes in the Svalbard archipelago, J. Geophys. Res., 84, 5620-5626, 1979.

Moreira, V., Seismotectonics of mainland Portugal and its adjacent region in the Atlantic (in Portugese, with English abstract), 29pp., Lisboa 1982.

Robson, G.R., K.G. Barr, and L.C. Luna, Extension failure: an earthquake mechanism, Nature, 218, 28-32, 1968.

Savostin, L.A., and A.M. Karasik, Recent plate tectonics of the Arctic Basin and of northeastern Asia, Tectonophysics, 74, 111-145, 1981.

Searle, R., The active part of Charlie-Gibbs fracture zone: A study using sonar and other geophysical techniques, J. Geophys. Res., 86, 243-262, 1981.

Sigvaldason, G.E., S. Steinthorsson, N. Óskarsson, and P. Imsland, Compositional variation in recent Icelandic tholeiites and the Kverkfjöll hot spot, Nature, 251, 579-582, 1974.

Solomon, S.C., Shear-wave attenuation and melting beneath the mid-Atlantic ridge, J. Geophys. Res., 78, 6044-6059, 1973.

Solomon, S.C., and B.R. Julian, Seismic constraints on ocean-ridge mantle structure: anomalous fault-plane solutions from first motions, Geophys. J. R. Astr. Soc., 38, 265-285, 1974.

Stauder, W., and G.A. Bollinger, The S-wave project for focal mechanism studies - earthquakes of 1962, Bull. Seismol. Soc. Am., 54, 2199-2208, 1964.

Stauder, W., and G.A. Bollinger, The S-wave project for focal mechanism studies, earthquakes of 1963, Bull. Seismol. Soc. Am., 56, 1362-1371, 1966.

Stefánsson, R., Methods of focal mechanism with application to two Atlantic earthquakes, Tectonophysics, 3, 209-243, 1966.

Sykes, L.R., Mechanism of earthquakes and nature of faulting on the mid-ocean ridges, J. Geophys. Res., 72, 2131-2153, 1967.

Sykes, L.R., Focal mechanism solutions for earthquakes along the world rift system, Bull. Seismol. Soc. Am., 60, 1749-1752, 1970.

Sykes, L.R., and M.L. Sbar, Focal mechanism solutions of intraplate earthquakes and stresses in the lithosphere; in: Geodynamics of Iceland and the North Atlantic Area (L. Kristjánsson ed.), D. Reidel Publishing Co., Dordrecht - Holland / Boston - USA, 1974.

Tams, E., Erdbeben im Gebiet der Nordenstkiöld See, Gerlands Beitr. z. Geophysik, 17, 325-331, 1927.

Trêhu, A.M., J.L. Nábělek, and S.C. Solomon, Source characterization of two Reykjanes Ridge earthquakes: Surface waves and moment tensors; P waveforms and non-orthogonal nodal planes, J. Geophys. Res., 86, 1701-1724, 1981.

Udias, A., A. Lopez Arroyo, and J. Mezcua, Seismotectonics of the Azores-Alboran region, Tectonophysics, 31, 259-289, 1976.

Vogt, P.R., Long wavelength gravity anomalies and intraplate seismicity, Earth Planet. Sci. Lett., 37, 465-475, 1978.

Ward, P.L., New interpretation of the geology of Iceland, Geol. Soc. Am. Bull., 82, 2991-3012, 1971.

Weidner, D.J., and K. Aki, Focal depth and mechanism of mid-ocean ridge earthquakes, J. Geophys. Res., 78, 1818-1831, 1973.

Whitmarsh, R.B., and A.S. Laughton, The fault pattern of a slow-spreading ridge near a fracture zone, Nature, 258, 509-510, 1975.

RECENT CRUSTAL MOVEMENTS IN CENTRAL EUROPE

Pavel Vyskočil

International Center on Recent Crustal Movements, 250 66 Zdiby, 98, Czechoslovakia

Abstract. In the following contribution the map of vertical crustal movements in the territory of Central Europe is presented. The map has been compiled using various levelling data available. In addition the main axes of strain field determined by geophysical as well as geodetic methods are shown. The features of movements are briefly described and it is concluded that the Alpine belt plays a dominant role in the tectonics of Central Europe.

Following the preliminary map of vertical movements in Central Europe, constructed by the International Center on Recent Crustal Movements (ICRCM) on the basis of stored data [Vyskočil, 1979], detailed work has recently been carried out to refine the map in the area of the Rhine Graben and along the border between France and the Federal Republic of Germany and Switzerland. Additional refinement was also made in the eastern half of Czechoslovakia, the West Ukraine, Hungary, and some parts of Roumania and Yugoslavia. The map of annual uplifts (vertical movements, Figure 1) was constructed in the same way as the preliminary map for the territory of German Democratic Republic, Poland and Czechoslovakia, published in 1968 [Vyskočil, 1969]. The content of this map is augmented by the main axes and horizontal directions of deformation based on geodetic data (the territory of German Democratic Republic and Czechoslovakia) [Thurm, 1977; Vyskočil, 1984] as well as on additional data derived from an analysis of the focal mechanism (the territory of the Federal Republic of Germany, France, Switzerland, Austria and Italy [Illies et al., 1971; Fourniquet et al., 1981].

Considering the plate tectonics viewpoint the presented map represents the dynamics of the border between the young orogens of the Alps and the Carpathians and old European platforms. There is a fairly pronounced elevation trending NW-SE; extending from the Rhine Massif and crossing variably the Bohemian Massif as far as its south - western margin. North of it lies the Labe fault line extending as a series of depressions towards the North Sea (Wilhelmshaven, Bremenhaven). To the south of the above mentioned elevated zone, runs a second depression belt that roughly follows the Alpine arc front but turns abruptly north-east in front of the West Carpathian arc. If one proceeds west-and south-westward from the Rhine Massif, one can observe a dominant tendency to uplift, but this is impossible to differentiate in greater detail because of the insufficient amount of data available from Belgium and France. Taken as whole, the Rhine Graben forms a longitudinal depression not exceeding a relative value of 1 mm/year and hence not perceptible on the map.

A rapid elevation change is a characteristic of the Alpine belt, but its differentiation still awaits a larger amount of data available. The Alpine high velocity arc is connected to a similar Carpathian arc by prominent subsidence covering the entire Vienna Basin and extending far southeast from Vienna. Carpathian uplift extends onto the Pannonian Lowland in Hungary. This uplift shows the Peripieniny lineament terminated by anomalous subsidence at its contact with the Vienna Basin. The closed subsidence, seen partly on our map south-east of Budapest, delineates the substantial part of the Pannonian Lowland bordered by the lower reach of the Tisa River and its confluence with Danube. Seashore from the Netherlands to Poland, shows a different tendency to subsidence.

The data on horizontal deformation shown, in Figure 1, are methodologically inhomogenous, but they nevertheless contribute to an understanding of the main deformation axes in the horizontal directions [Fourniquet et al., 1981; Illies et al., 1981]. In this respect it is thus possible to use these data to infer the main directions of horizontal compression which runs perpendicular to the Alpine arc. This tendency is discernible in immediate vicinity of the Alpine foothills and in southern Bohemia. This tendency might carry as far as the border between German Democratic Republic and Czechoslovakia [Thurm, 1977], due to the transmission of the Alpine compression through the Bohemian Massif. The axes of horizontal deformation on the outer West Carpathian arc, based on geodetic data used by the author [Vyskočil,

Copyright 1987 by the American Geophysical Union.

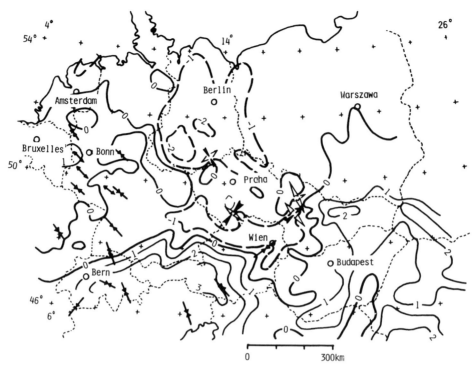

Fig. 1. Annual uplifts and horizontal deformation in central Europe.

1984], are directionally akin to the compression axis at the south-western margin of the Vienna Basin. As regards their origin, however, they seem to be largely controlled by an own, more or less independent movement of the Carpathian arc. The tendency to extend in the direction normal to the Carpathian arc may serve as supporting evidence for this hypothesis. Moreover, the similar results were published as it is to be seen [Somov et al., 1975] from the eastern part of the Carpathian arc [Vyskočil, 1984].

Independent transverse dynamics effects (spreading tendency) can probably be seen also in the Rhine Graben and Rhine Massif, but geodetic data on horizontal deformation are still lacking there. From the predominant compression in the Rhine Graben it is possible to infer an extension in the SW-NE direction characterizing the graben as a rift zone. In order to determine what horizontal deformations are taking, than it is now necessary to have a larger amount of homogeneous data, particularly those aquired by geodetic methods designs specifically detect strain.

Considering the results described in this contribution, it is possible to arrive at the conclusion that recent dynamics conforms to the principal tectonic features of Central Europe. Recent movements clearly reflect the Alpine and Carpathian arcs, including their foreland, and partly the Pannonian Basin as well. In the recent movement pattern it is also possible to detect the main tendencies of dynamics and positions of principal fault lines over the area under study [Zátopek et al., 1975]. Judging from the directions of the main compression axes in the southern part of the Bohemian Massif and in the foreland of Alps, the possibility cannot be ruled out that the Alpine stress acts on the Bohemian Massif even now [Zátopek, 1980]. Generally speaking, the pattern presented in this paper is of a schematic nature; further refinements will be possible, when more data become available. Nevertheless even, repeated geodetic measurement results available now make an essential contribution to a better understanding of the principal features in present-day tectonics of Central Europe.

References

Fourniguet, J., J. Vogt, and C. Weber, Seismicity and recent crustal movements in France. Tectonophysics, 71, 195-216, 1981.

Illies, H., H. Baumann, and B. Hoffers, Stress pattern and strain release in Alpine foreland. Tectonophysics, 71, 157-172, 1981.

Somov, V. I., and V. G. Kuznecova, Results of geodetic and geophysical investigations of recent crustal movements in the Soviet part of eastern Carpathians. Tectonophysics, 29, 377-382, 1975.

Thurm, H., and P. Bankwitz, Rezente horizontale Deformationen der Erdkruste in Suedostteil der GDR. Petersmans Geogr. Mitt., H.4, Berlin, 281-304, 1977.

Vyskočil, P., A comparison of preliminary maps of annual velocities of vertical crustal movements on the territory of the German Democratic

Republic, of the Polish People's Republic and of the Czechoslovak Socialist Republic. Problems of ~ent Crustal Movements. <u>Third Intern. Symp. of Cn..., Leningrad 1968</u>, AN SSSR, Moscow, 93-100, 1969.

Vyskočil, P., Heat flow, crustal thickness and recent vertical movements. <u>Terrestrial Heat Flow in Europe</u>, Springer Verlag, Berlin-Heidelberg, 119-125, 1979.

Vyskočil, P., Results of recent crustal movement studies. <u>Transaction of the ČSAV, ser. Mathem. & Natur.Sci.</u>, <u>92</u>, No.8, Academia, Praha, 103, 1984.

Zátopek, A., and B. Beránek, Geophysical syntheses and crustal structure in central Europe. <u>Studia Geophys.et Geod.</u>, Academia, Praha, <u>19</u>, 121-133, 1975.

Zátopek, A., Východoalpská zemětřesení a seismicita okrajů Českého Masivu. in Czech., Proc."<u>Výzkum hlubinné geologické stavby Československa</u>", Loučná, 1980, Geofyzika n.p. Brno, 79-82, 1980.

LITHOSPHERIC DEFORMATION DEDUCED FROM ANCIENT SHORELINES

P.A. Pirazzoli

CNRS-Intergéo, 191 rue Saint-Jacques, 75005 Paris, France

D.R. Grant

INQUA Shorelines Commission, 5 Birchview Court, Nepean K2G 3M7, Canada

Abstract. A review of some shoreline studies which have contributed to an understanding of lithospheric deformation.

Introduction

Former shorelines, like their modern counterparts, represent the position of gravitational equilibrium between lithosphere and hydrosphere. In tectonically quiescent areas, sea level is considered to be eustatic and until recently the aim of IGCP Project 61 (1974-82) was to document its recent history by means of "a graph of the trend of mean sea level during the last deglaciation and up to the present time."

However, most published sea-level curves differ markedly from one another (Bloom, 1977; Nivmer Information, 1978-82). The local and regional variations support the conclusions reached by several authors (Walcott, 1972; Mörner, 1976; Pirazzoli, 1976; Clark et al., 1978; Newman et al., 1979) that no region could be considered absolutely stable, in comparison with all other areas, for the purpose of inferring crustal movements. Hence the goal of IGCP-61 was effectively impossible to attain. As summarized by Faure (1981), following the general agreement that began emerging after the 26th International Geological Congress (1980), "the new goal is to define the history of local or regional sea levels." The rationale is simply that precise and reliable local sea level trends are the sum of all tectonic, eustatic and gravitational factors, and thus offer the only means of extracting information on vertical movements of the lithosphere.

Primarily by using the disappearance of continental ice sheets and the redistribution of water in the world ocean, global isostatic models of the sea-level response to changing earth loads were developed (Walcott, 1972; Clark et al., 1978). The models enabled theoretical local sea-level histories to be predicted which, by comparison with measured sea-level curves, permitted the model to be further refined. At present, most of the observed regional variations are explained by the model but as it is based on a relatively few and unevenly distributed control points, it is considered a first approximation that needs validation at many places where no data exists.

The value of the model for lithospheric distortions is that, by defining sea level changes due solely to deglacial meltwater redistribution, any departures of observed sea level are likely attributable to crustal movements. In the same way, rheological properties and mantle activity can be deduced, at least in part, from the delevelling of a given age strandline.

Concurrently, satellite ranging has revealed that the topography of the geoid or sea level surface (Gaposchkin, 1973; Marsh and Martin, 1982) has a relief of up to 200 m. This led Morner (1976) to suggest that mass changes within the mantle could be deduced from variations in the height of a paleoshore, and from regional transgressions and regressions.

The delevelling of shorelines thus results from a variety of independent motions. Sea level studies therefore can provide a wealth of information pertaining to internal and external effects on the lithosphere. The following is a review of some shoreline studies which have contributed to an understanding of lithospheric deformation.

Deformation of Glacio- and Hydro-Isostatic Origin

The growth and decay of Pleistocene ice sheets, with their resulting isostatic crustal deflections and attendant distribution of

Copyright 1987 by the American Geophysical Union.

subcrustal mass and ocean water, is perhaps the grandest natural experiment in lithospheric deformation that could be devised. Global in its many side effects, it approximates even exceeds, the growth of a mountain range or the filling of sedimentary basin in terms of the mass involved, and the speed of deflection. In short, delevelled shorelines are eloquent testimony to the outer structure of the Earth, and its behaviour under stress.

Little wonder then that for over 100 years, the pattern of upwarped shorelines in glacial districts has been studied as an elegant expression of Earth's response to the transient loads of ice and water. In the last 50 years growing interest in the discrete properties of crust and mantle has focused on areas of rebound. From these the first estimates of crustal rigidity (the flexural parameter), the viscosity of the upper mantle, and relaxation time (half life) of the process have come from an analysis of dated delevelled shorelines, primarily in the central areas of the ice sheets: Gulf of Bothnia for Scandinavian Ice Sheet, and Hudson Bay for the Laurentide Ice Sheet. The hypothesis of a peripheral bulge built of mantle material squeezed from beneath the flooded regions (Daly, 1934) was tested in the ice-marginal area of eastern North America. Conversely, local hydro-isostatic phenomena such as the crustal subsidence due to the water load of the large pluvial Lake Bonneville was evaluated from the updomed shorelines and found to be as predicted from existing theory (Crittenden, 1963). As a result, crustal adjustments under all large man-made water reservoirs are now routinely calculated and monitored by means of precise relevelling. Studies of glacial isostasy in several regions had thus succeeded in defining some of the dynamical factors (e.g. Andrews, 1970) and these appeared to adequately explain observable effects within small areas of study. However, various regions gave differents results which suggested that some factors had been overlooked or over-generalized, and that a new approach was needed.

It was the advent of the IGCP Sea level Projects that served to bring together experts from a variety of disciplines to attempt a global solution. The results of the first such meeting in 1977 culminated in the book Earth Rheology, Isostasy and Eustasy (Mörner, 1980), which sparked a renewed interest in assembling data on delevelled shorelines in order to reveal fundamental characteristics and behaviour of the Earth's surface layers. Using modern highspeed computers, and elegant mathematical models of gravitationally self-consistent Earth, it became possible to simulate the total global response of the shrinking ice sheets. In this way the four main factors could be accommodated simultaneously: deglacial upwarp, subcrustal transfer from extraglacial tracts, meltwater accession, and hydroisostatic depression. The position of world sea level on the continents was simply the net result of these independent adjustments. Thus, with observed sea-level histories (RSL curves) as control, the model could be reiterated so as to adjust the crust and mantle parameters until cause and effect were congruent. (Cathles, 1975; Peltier and Andrews, 1976; Clark, 1980). By this means the sea-level change, or more particularly the crustal movement of any point on the surface of the Earth, as a response to the shrinking of a complex of ice masses, could be calculated. Five broad zones of differential movement, as a function of distance from the glaciated regions, have been delineated. For the first time it became clear why the rate, direction and magnitude of crustal movement varied so much from place to place. Not only did these models succeed in duplicating reasonably well, the observed changes, they served to explain and reconcile the disparate movements in seemingly comparable areas (e.g. the emergence of the South Pacific Islands versus the submergence of Micronesia).

Differences between predicted and observed deformation are presumed to be due to regional differences in crustal/rheological properties such as crustal thickness and temperature, and mantle viscosity profile, compared to the global averages used in the models. The discrepancies have prompted theoreticians to propose various configurations of mantle viscosity and circulation. Concurrently, practical field trials are being carried out in certains areas (the Canadian Atlantic Provinces, for example) to investigate the response of crust and sea level to more detailed local (ice) load changes (Quinlan and Beaumont, 1982). A program of sea-level studies is underway there to refine the model, and to solve the equally interesting inverse problem of reconstructing the vanished load from its legacy of crustal recovery.

The displacement of former marine and lake shorelines in response to glacial unloading and to water loading has thus given crucial information on lithospheric deformation both on plate margins and interiors. Transects perpendicular to continental margins have shown various degrees of continental flexuring, resulting from the depression of the ocean floor by meltwater accession and from changes in the rigidity of the lithosphere, in Brazil (Martin et al., 1979-80), in Senegal (Faure et al., 1980) and in Australia (Chappell et al., 1982; Hopley, 1982).

Field studies are now focused on crucial and deficient areas with a view to precise definition of the progress of crustal movement. Analysis can now integrate all factors in global solutions, and can reliably predict consequences of hypothetical changes. Both approaches are thus seen as essentiel components of any international rheological program.

Thermo- and Volcano-Isostatic Deformation

It is now generally accepted that as the ocean crust spreads away from its point of origin along submarine ridges, it cools and thickens, thereby increasing its density. As a result the seafloor subsides isostatically thus gradually submerging oceanic islands as they are carried laterally. In tropical waters the subsidence is increased by the load of coral reefs which thicken to maintain their sea level position.

Where the oceanic crust approaches a "hot spot" the normal cooling process is reversed, so that the crust is heated, becomes less dense and thinner, and thus rises. During transport away from the hot spot there is renewed cooling and subsidence (Menard, 1973; Detrick and Crough, 1978). Cycles of repeated upheaval and sinking on Pacific atolls can therefore be interpreted as evidence of plate movement in relation to hot spots (Coudray and Montaggioni, 1982; Scott and Rotondo, 1983).

Near hot spots, extrusion of lavas often occurs. The resulting load produces isostatic effects comparable to those of an ice sheet with equivalent mass; namely a depression under the volcanic pile and a peripheral raised rim (Walcott, 1970b).

These deformations can be documented in detail by mapping shorelines, either as a sequence at one place by means of drilling in coral reefs, or by correlation of a given level along island chains so as to construct a mid-ocean profile. The former method has given average subsidence rates of 0.2 mm/y since 60 Ma in the Marshall Islands (Menard and Ladd, 1963) and 0.12 mm/y since the Pliocene in the Mururoa Atoll (Labeyrie et al., 1969). The latter has given information on vertical movements along intraplate transects which has helped refine global isostatic models, such as those by Walcott (1972) and Clark et al. (1978).

Where the translation movement of an oceanic plate over a hot spot has produced a line of islands, interaction between isostatically depressed areas and uplifted rims of nearby islands is possible and may produce complicated sequences of vertical deformation (Scott and Rotondo, 1983).

Verifications of these models have been attempted in the Hawaii and Society Islands. In Hawaii, the occurence of active subsidence has been deduced from tide gauge data, while bathymetric maps show that a 600 ± 200 m high arch has developed around the island (Walcott, 1970). On Kawai Island, on the other hand, apparent uplift has been ascribed to the position of this island on the arch of Hawaii (Moore, 1971).

In the Society Islands, a slight (0.5 m) but progressive decrease in the emergence of the 3000-yr BP shoreline is observed from Maupiti to Bora Bora, Raiatea, Huahine, Moorea and Tahiti. Here, the predominant mass of Tahiti is assumed to cause an active flexuring of the lithosphere (Pirazzoli, 1983).

The elevation of the isostatically raised rim around Tahiti is probably less than 100 m above the surrounding ocean floor, since bathymetric maps with 100 m depth intervals (Monti and Pautot, 1974) do not show any remarkable feature. The raised atoll of Makatea, however, 245 km NE of Tahiti, reaching some 110 m in altitude, is likely to be located right in the middle of the rim. The limestone cliffs of Makatea are cut by three lines of notches: 1.0-1.5 m, 5-8 m and 20-25 m in elevation (Montaggioni et al., in press). These notches correspond to the sea-level positions at the time of the maximum stand in the Holocene, and of two late Pleistocene stands, dated 100,000 to 140,000 yr and more than 200,000 yr by Veeh (1966).

The K-Ar ages of the Tahiti eruptions vary from 0.16 to 0.96 Ma (Dymond, 1975) and underwater eruptions are known to occur at present not far from the island, between Tahiti and Mehetia (Okal et al., 1980). This means that Makatea's uplift is contemporary with the magmatic outflows of the Tahiti hot spot.

The outer limit of the uplifted rim of Makatea reaches as far as the first NW Tuamotu atolls, as shown by the end of the traces of elevated reefs, forming an approximate arc of a circle at about 300 km from Tahiti (McNutt and Menard, 1978; Pirazzoli, 1985).

Crustal Block Movements From Subduction Zones

Subduction, collision and sliding processes, in areas of lithospheric plate convergence, often produce important crustal movements which are recorded by ancient shorelines. Investigation and dating of former shorelines may be used therefore to determine the age, distribution and succession of the vertical displacements, as well as the limits between lithospheric blocks of different tectonic behaviour. In some cases, it is even possible to predict probable areas of impending great magnitude earthquakes and the vertical crustal movements which will probably accompany these earthquakes.

In the peninsulas of southwestern Japan, late Quaternary spasmodic uplifts have left superimposed marine terraces, some of which are Holocene in age. Recurrence times of sudden vertical displacements range here from 90-264 yr (Yonekura, 1975) to 1,000-2,000 yr (Matsuda et al., 1978; Nakata et al., 1979; Yoshikawa et al., 1981). The movements generally displace crustal blocks some dozen kilometers wide; it has been predicted that the next great earthquake in southwest Japan will occur in the Tokai district (Ando, 1975; Yonekura, 1975).

In Southern Greece, an independently active block of lithosphere approx. 150 km long, including the Levka Mountains in Crete, and Antikithira Island, has been identified from shorelevel studies. This block has experienced paroxysmal tectonics probably related to the subduction of the Ionian slab beneath the Hellenic Arc (Le Pichon and Angelier, 1979): ten rapid subsidence movements (from 10 to 25 cm each time) without noticeable tilting, between 4,000 and 1,700 yr BP; then, about $1,530 \pm 40$ yr BP, a conspicuous 10-m uplift, with northeastward tilting, in a single event (Thommeret et al., 1981; Pirazzoli et al., 1982).

Many local measurements of vertical rates of deformation have been obtained from late Quaternary shorelines in active tectonic areas; it is not possible to mention all of them here and more detailed references can be found in the lists of publications of the IGCP-Projects 61 and 200 and of the INQUA Commissions on Shorelines and on Neotectonics. What should be emphasized here is that in zones of plate convergence tectonic stresses operate on areas which are usually much larger than those showing crustal deformation. As a matter of fact, vertical displacement during great earthquakes occur in one or more crustal blocks, depending on the position of the epicentres. The movements of each block certainly reflect on the nearby blocks, but the latter are often not displaced at the same moment, while more distant crustal blocks may be so. The following two examples will enlighten this point.

In Eastern Crete and the island of Rhodes, tectonic movements differ in type and in time from those already described for the 150-km long lithospheric block in western Crete. More to the east, however, an age very similar to that obtained in western Crete ($1,545 \pm 40$ yr BP) has been found for the exceptional uplift of extensive coastal benches on the Turkish coast, near Alanya (Kelletat and Kayan, 1983).

In the Ruykyu Islands, Japan, the age and level of the Holocene emerged shorelines vary from island to island and sometimes from one part to the other of the same island (Pirazzoli, 1978; Kawana and Nishida, 1980; Koba et al., 1982). However, exceptional movements of sudden uplift have occured at the same time (ca. 2,350 yr BP) in south Okinawa Island (Kawana and Pirazzoli, 1983) and, 400 km away, in the Tokara and Kodakara Islands (Koba et al., 1982), without any evidence of contemporaneous movements in other islands located between these two uplifted areas.

This seems to indicate that unusual vertical deformation is likely to be the result of unusual local stresses. A possible interpretation is that irregularities on the surface of the sinking slab occasionally hook, in one or more places at the same time, on to blocks of lithosphere in the insular arc. These events may interrupt the smooth progression of the slab thus producing in some cases tectonic paroxysmal phenomena. As the constraints increase sufficiently, however, faults occur. Part of the slab is suddenly subducted while the overthrusting lithospheric blocks, suddenly freed, are uplifted and tilted whereas other crustal blocks glide over the slab without showing any vertical deformation.

Conclusions and Recommendations

Investigation of ancient shorelines is a very effective means of improving the understanding of the rheological properties of the Earth and of recent lithospheric deformation. New and existing sea-level data should be incorporated in global and regional models of geoid height. By computer simulation, these models should allow extrapolation and prediction of vertical deformation within poorly known areas.

However, data have only limited value if they are too isolated in space and in time; on the other hand, a good global coverage of the oceans would need tens of thousand of reliable data. A realistic approach would be to intensify the investigation along a number of well-chosen transects in key areas providing diagnostic evidence to evaluate assumptions underlying any models which may be developed. The main aim is to reconstruct accurate sequences of palaeogeoidal profiles.

Transects should be perpendicular to ice masses, to continental margins and to active plate boundaries. Changes in elevation of shorelines located near an active trench on the subducting side can even be used to measure rates of plate motion (Nakamura, 1982). Other transects should cross areas of presumed anomalies in the lithosphere and upper mantle (hot spots) and should follow island chains.

The most useful results will be obtained in those areas where sequences of dated shorelines can be obtained in the same place. Rapidly uplifting areas showing several emerged superimposed shorelines are especially worthy of thorough study (Mesolella et al., 1969; Chappell, 1974; Hillaire Marcel and Fairbridge, 1978; Pirazzoli et al., 1982). Very useful results can be obtained also by drilling in sedimentary sequences where datable material (coral reefs, peat layers, etc.) is expected at various depths.

In active tectonic areas, tide-gauge data will provide valuable information on the most recent crustal deformation, which, when compared with the trends indicated by the ancient shorelines, might be of greatly different rate or direction than longer-term geological movement, and thus might possibly forewarn of stress and possible seismicity (Riddihough, 1983).

References

Ando, M., Source mechanism and tectonic significance of historical earthquakes along the Nankai Trough, Japan, Tectonophysics, 27, 119-140, 1975.

Andrews, J.T., A geomorphological study of postglacial uplift with particular reference to Arctic Canada, Institute of British Geographers, Spec. Publ., 2, 156 pp., 1970.

Bloom, A.L., Atlas of sea-level curves, IGCP Project 61, Cornell Univ., 1977.

Cathles, L., The viscosity of the Earth's mantle, Princeton Univ. Press, Princeton NJ, 386 pp., 1975.

Chappell, J. Geology of coral terraces, Huon Peninsula, New Guinea: a study of Quaternary tectonic movements and sea-level changes, Geol. Soc. Am. Bull., 85, 4, 553-570, 1974.

Chappell, J., Rhodes, E.G., Thom, B.G., and Wallensky, E., Hydro-isostasy and sea-level isobase of 5500 B.P. in North Queensland, Australia, Marine Geology, 49, 81-90, 1982.

Clark, J.A., A numerical model of worldwide sea-level changes on a visco-elastic earth, in N.A. Mörner (editor): Earth Rheology, Isostasy and Eustasy, J. Wiley & Sons, 599 pp., 1980.

Clark, J.A., Farrell, W.E., and Peltier, W.R., Global changes in postglacial sea level: a numerical calculation, Quatern. Res., 9, 265-287, 1978.

Coudray, J., and Montaggioni, L., Coraux et récifs coralliens de la province indo-pacifique: répartition géographique et altitudinale en relation avec la tectonique globale, Bull. Soc. Géol. France, 24, 981-993, 1982.

Crittenden, M.D. New data on the isostatic deformation of Lake Bonneville, U.S. Geol. Surv., Prof. Pap., 454-E, 31 pp., 1963.

Daly, R.A., The changing world of the Ice Age, Yale Univ. Press., New Haven, 271 pp., 1934.

Detrick, R.S., and Crough, S.T., Island subsidence, hot spots and lithospheric thinning, J. Geophys. Res., 83, 133, 1236-1244, 1978.

Dymond, J., K-Ar ages of Tahiti and Moorea, Society Islands, and implications for the hot-spot model, Geology, 3, 236-240, 1975.

Faure, H., Extraits du rapport de la Section S.08 Quaternaire et Géomorphologie, Nivmer Information, 7, 9.6, 1981.

Faure, H., Fontes, J.C., Hebrard, L., Monteillet, J., and Pirazzoli, P.A., Geoidal change and shore-level tilt along Holocene estuaries: Sénégal River area, West Africa, Science, 210, 421-423, 1980.

Gaposchkin, E.M., 1973 Smithsonian standard Earth, SAO Special Report, 353, Smithsonian Inst., Astrophys. Observ., 1973.

Hopley, D., The geomorphology of the Great Barrier reef, Wiley & Sons, 453 pp., 1982.

Kawana, T., and Nishida, H., Preliminary report on the formative periods of the notches of Yoron, Okinawa, Miyako and Ishigaki islands, the Ryukyu Islands, South Japan (in Japanese with English abstract), Geol. Stud. Ryukyu Is., 5, 103-123, 1980.

Kawana, T., and Pirazzoli, P.A., Late crustal movements in Okinawa Islands, the Ryukyus, Japan, Intern. Symp. Coastal evolution in the Holocene, Tokyo, Aug. 29-31, Abstracts, 53-56, 1983.

Kelletat, D., and Kayan, I., First C-14 datings and late Holocene tectonic events on the Mediterranean coastline, west of Alanya, Southern Turkey (in Turkish, with English abstract), Bull. Geol. Soc. Turkey, 26, 83-87, 1983.

Koba, M., Nakata, T., and Takahashi, T., Late Holocene eustatic sea-level changes deduced from geomorphological features and their 14-C dates in the Ryukyu Islands, Japan, Palaeogeogr., Palaeoclim., Palaeoecol., 39, 231-260, 1982.

Labeyrie, J., Lalou, C., and Delibrias, G., Etude des transgressions marines sur l'atoll de Mururoa par la datation des différents niveaux de corail, Cah. Pacif., 13, 59-68, 1969.

Le Pichon, X., and Angelier, J., The Hellenic arc and Trench system: a key to the neotectonic evolution of the eastern Mediterranean area. Tectonophysics, 60, 1-42, 1979.

Marsh, J.G., and Martin, T.V., The Seasat altimeter mean sea surface model, J. Geophys. Res., 87, C5, 3269-3280, 1982.

Martin, L., Suguio, K., Flexor, J.M., Bittencourt, A., and Vilas-Boas, G., Le Quaternaire marin brésilien (littoral pauliste, sud fluminense et bahianais), Cah. ORSTOM, sér. Géol., 11, 1, 95-124, 1979-1980.

Matsuda, T., Ota, Y., Ando, M., and Yonekura, N., Fault mechanism and recurrence time of major earthquakes in southern Kanto district, Japan, as deduced from coastal terrace data, Geol. Soc. Am. Bull., 89, 1610-1618, 1978.

McNutt, M., and Menard, H.W. Lithospheric flexure and uplifted atolls, J. Geophys. Res., 83, B3, 1206-1212, 1978.

Menard, H.W., Depth anomalies and the bobbing motion of drifting islands, J. Geophys. Res, 78, 23, 5128-5137, 1973.

Menard, H.W., and Ladd, H.P., Oceanic islands, sea-mounts, guyots and atolls, in A.E. Maxwell (editor): The Sea, New York, Intersci., 365-387, 1963.

Mesolella, K.J., Matthews, R.K., Broeker, W.S. and Thurber, D.L., The astronomical theory of climatic change: Barbados data, J. Geol., 77, 250-274, 1969.

Montaggioni, L.F., Richard, G., et al., Aspects of the geology and marine biology of Makatea, an uplifted atoll, Tuamotu Archipelago, Central Pacific Ocean, Litoralia, in press.

Monti, S., and Pautot, G., Bathymetrie Pacifique sud, Carte Tahiti 1/1,000,000, Centre Océanologique de Bretagne, 1974.

Moore, J.G., Relationship between subsidence and volcanic load, Hawaii, Bull. Volcanologique, 34, 562-576, 1971.

Mörner, N.A., Eustasy and geoid changes, J. Geol., 84, 123-151, 1976.

Mörner, N.A., (editor), Earth rheology, isostasy and eustasy, John Wiley & Sons, 599 pp., 1980.

Nakamura, K., Plate motion may be measured in situ as subsidence at trenches, J. Geol. Soc. Japan, 28, 3, 174-176, 1982.

Nakata, T., Koba, M. et al., Holocene marine terraces and seismic crustal movements, Sci. Rep. Tohoku Univ., Geogr., 29, 2, 195-204, 1979.

Newman, W.S., Marcus, L.F., and Pardi, R.R., Palaeogeodesy: late Quaternary geoidal configurations as determined by ancient sea levels, IAHS Publ., 131, 263-275, 1979.

Nivmer Information, n° 2, 3, 4, 5, 7, 8, Montrouge, 1978-1982.

Okal, E.A., Talandier, J., Sverdrup, K.A., and Jordan, T.H., Seismicity and tectonic stress in the south-central Pacific, J. Geophys. Res., 85, B 11, 6479-6495, 1980.

Peltier, W.R. and Andrews, J.T., Glacial isostatic adjustment. I: The forward problem, Geophys. J.R. Astron. Soc., 46, 605-646, 1976.

Pirazzoli, P.A., Les variations du niveau marin depuis 2000 ans, Mem. Lab. Geomorphol. Ec. Prat. Hautes Et., 30, 421 pp., 1976.

Pirazzoli, P.A., High stands of Holocene sea levels in the northwest Pacific, Quatern. Res., 10, 1-29, 1978.

Pirazzoli, P.A., Mise en évidence d'un flexure active de la lithosphère dans l'archipel de la Société (Polynésie française), d'après la position des rivages de la fin de l'Holocène, Comptes Rendus Acad. Sci. Paris, II, 296, 695-698, 1983.

Pirazzoli, P.A., Thommeret, J.&Y., Laborel, J., and Montaggioni, L.F., Crustal block movements from Holocene shorelines: Crete and Antikythira (Greece), in X. Le Pichon, S.S., Augustithis and J. Mascle (editors): Geodynamics of the Hellenic Arc and Trench, Tectonophysics, 86, 27-43, 1982.

Pirazzoli, P.A., Les anciens rivages, témoins des déformations de la lithosphère, Bull. Soc. Geol. France, 3, 343-351, 1985.

Quinlan, G., and Beaumont, C., The deglaciation of Atlantic Canada as reconstructed from the postglacial relative sea-level record, Canad. J. Earth Sci., 19, 2232-2246, 1983.

Riddihough, R., Contemporary vertical movements and tectonics on Canada's west coast (abstract), Geol. Ass. Canada, Program, 8, 457, 1983.

Scott, G.A.J., and Rotondo, G.M., A model to explain the differences between Pacific plate island-atoll types, Coral Reefs, 1, 139-150, 1983.

Thommeret, Y.&J., Laborel, J., Montaggioni, L.F., and Pirazzoli, P.A., Late Holocene shoreline changes and seismo-tectonic displacements in western Crete (Greece), Zeits. Geomorphol., Suppl.-Bd., 40, 127-149, 1981.

Veeh, H.H., Th-230/U-238 and U-234/U-238 ages of Pleistocene high sea level stand, J. Geophys. Res., 71, 14, 3379-3386, 1966.

Walcott, R.I., Flexure of the lithosphere at Hawaii, Tectonophysics, 9, 435-446, 1970a.

Walcott, R.I., Flexural rigidity, thickness, and viscosity of the lithosphere, J. Geophys. Res., 75, 20, 3941-3954, 1970b.

Walcott, R.I., Past sea levels, eustasy and deformation of the Earth, Quatern. Res., 2, 1-14, 1972.

Yonekura, N., Quaternary tectonic movements in the outer arc of southwest Japan, with special reference to seismic crustal deformation, Bull. Dept. Geogr. Univ. Tokyo, 7, 19-71, 1975.

Yoshikawa, T., Kaizuka, S., and Ota, Y., The landforms of Japan, Univ. Tokyo Press, 222 pp., 1981.

RECENT CRUSTAL MOVEMENTS AND GRAVITY IN ARGENTINA: A REVIEW

Antonio Introcaso

Instituto de Física de Rosario (IFIR), Avda. Pellegrini 250, 2000 Rosario, Argentina

Abstract. In order to find probable vertical movements, gravimetrical models and gravity changes along Argentina's EW profiles were analyzed. In this work, a general tendency of gravity decrease towards the Andes, reaching approximately -1.5 mGal, was found along the central profile. Two more EW profiles - one in the north and the other in the south of Argentina were also remeasured. A careful investigation was performed on the Caucete or Pie de Palo, 1977, earthquake and a gravity difference of -0.37 mGal was found by comparing the measurements made before and after the seismic event. An assumed level variation of +1.30 m and the vertical gradient of gravity known from observation explain well a consistent relationship between the gravity decrease and the elevation of Pie de Palo's East area after the 1977 earthquake (reverse faulting after focal mechanism study). At the same time, studies performed from 1949 to 1982 in Caucete area or SE Pie de Palo yielded a gravity difference of -0.25 mGal with respect to near stations. The great sedimentary basins of Salado and Colorado show a thick sedimentary development (about 7000 m). Both basins show positive gravity anomalies that increase significantly when the upper crust is normalized. The formation of the basin probably develops when asthenospheric materials arise and provoke stresses in the crust which are then released by plastic flow in the lower part and by faulting in the brittle upper part. Then subsidence is controlled by cooling and sedimentary loading. The present high gravity will tend to decrease in accordance with the present active subsidence and the ulterior loading.

Introduction

This report refers to the studies made on various areas in Argentina by various investigators. Together with other sorts of data, these studies discuss "g" variations in time, which are related with cortical movements and deformations, although the grade of consistency looks different in each case.

Copyright 1987 by the American Geophysical Union.

Luckily, measurements of gravity in Argentina were made 30 or more years ago with much care, in such a looper-motion scheme of repeating station readings as A-B-A-B-C... . Thus, these data can be compared with the "g" values, remeasured in recent years with modern high-sensitivity gravimeters, to a reasonable degree of reliability.

In this report, we will review temporal gravity variations in several areas of Argentina, seeking for variety of aspects, they are, the Andean cordillera, the Pampean hills, and the Pampean plains which include the great Bonearense sedimentary basins.

Unluckily, geological, geophysical and geodesic information, which would supplement interpretation of "g" variations, is insufficient in many cases, for example the Northern EW profile. This is not the case of Caucete earthquake (San Juan) of November 1977, where we can use data of precise measurements of "g", relevelling, geological studies of the field (and by photogeology), seismic studies, etc.

Crustal Movements and Gravity Changes in the Epicentral Area of Caucete Earthquake on November 23, 1977

This widely known earthquake has been studied by Volponi [1979], Triep [1979], and Cardinalli [1984], and others.

It very seldom happens that we make precise surveys of a future seismic area prior to an earthquake. Regarding the Caucete earthquake on November 23, 1977, however, a precise gravimetric survey [Introcaso and Huerta, 1972] and other three high precision surveys had been performed in the epicentral area prior to this strong earthquake. This earthquake is located at the Pie de Palo hill which is one of the Pampean hills formed by infra Cambrian rocks limited by compression faults originated or reactivated by the Andean orogeny. It has also been recognized that they are subject to significant rising radial movements.

Relevelling and gravity remeasurements were performed after this important earthquake.

The faulting in the Caucete or Pie de Palo earthquake, as well as that in the Las Lajas

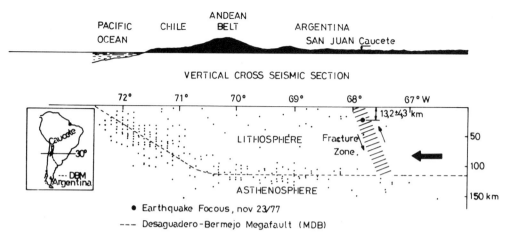

Fig. 1. Location of the Caucete earthquake of November 23, 1977 [Volponi, 1978]. Approximate latitude is 31°S.

earthquake in 1944, corresponds to a reverse fault of compression type. The oriental plane slides over the occidental plane which plunges mainly to the east (Fig. 1, Volponi [1979]) for about 70 in angle.

Later, Triep and Cardinalli [1984, in print] used the first movements of P wave and obtained solutions of focal mechanisms for 50 not very deep earthquakes (h < 60 km) in the adjacency of Desaguadero-Bermejo Megafault (MDB, see Figure 1 for its location) and in the neighbouring areas between 29° and 30°S in latitude. Most of them belong to a sequence of events that started with the earthquake of November 1977. It was estimated that the event included three shocks, with the second one being most energetic of all.

With only few exceptions, the mechanisms are of reverse faulting type. Both of the nodal planes in each solution correlate, in general, with geo-lignment which are mostly supposed to have existed before the earthquake (some of them are recognizable on the Landsat images).

Gravimetrical Measurements in the Epicentral Area

Introcaso [1972] made a gravimetrical survey all along an EW profile running from the Atlantic Ocean to the Pacific Ocean, in the proximity of the 32 parallel. This geotransverse crosses the epicentral area of the Caucete, 1977, earthquake.

Not counting the effect of regional gravity that pertains to the Andine structure, there exists a residual Bouguer anomaly of ca.+50 mGal over the Pie de Palo hill which indicates a sense just opposite to what was expected (direct correlation between topography and gravity).

This is similar to the case of the Big Horn Mountains [Wollard, 1959]. Because of Pie de Palo's restricted width (about 30 km), we may presume that the cortical thickness of the area will support the excess mass of the hill and no isostatic reactions will originate. Figure 2 shows a part of the performed levellings and indicates that the oriental block is significantly uplifted relative to the occidental block. If we

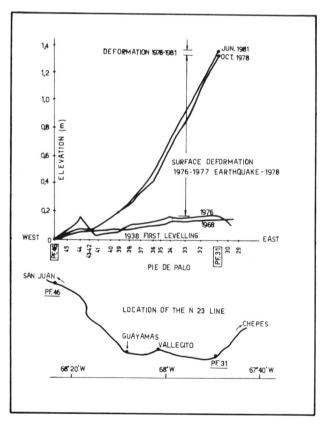

Fig. 2. Location of the 23 N line. The different levellings along it show uplift of the east block relative to the west block (see Volponi et al. [1982]).

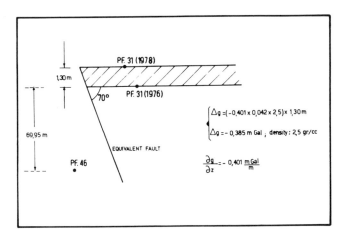

Fig. 3. Scheme for theoretical calculation of gravity change associated with the 1977 earthquake [Volponi et al,, 1982].

assume vertical movement for +1.30 m between 1970 and 1978, then gravity change would be calculated in such a manner as given in Figure 3 [Volponi et al., 1982].

In spite of the unfavourable measurement conditions (lack of simultaneity between levelling and "g" measurement; measurement of very small values, etc.), a difference of -0.370 mGal was observed. This data looks reasonable in reference to the gravity variation, -0.385 mGal calculated for the elevation change of 1.30 m.

The important fact is that the forces that caused the Caucete earthquake were acting since the preseismic time, producing significant deformations on the surface. They are still working after the earthquake. The direct correlation between topography and gravity in Pie de Palo, which we have already pointed out, and the uplifting movements in these days indicate the tectonic forces of anti-isostatic mode.

Crustal Movements along the San Juan - Mendoza Meridian Line

In 1982 the Instituto Sismológico Zonda (ISZ) releveled the San Juan - Mendoza N 24 line (157 km) considering Chepes as vertically fixed (a supposedly stable place located on the neighbouring province of La Rioja). The first levelling was performed in 1938 by the Instituto Geografico Militar Argentino (IGM). Comparison of the values revealed land subsidence at all the stations. Thus the terminal stations of San Juan and Mendoza subsided for 39.24 and 63.63 cm, respectively, relative to Chepes [Sisterna and others, 1982]. Settlement of alluvial ground, variation in depth of water table and tectonic movements were pointed out as the causes of the subsidence. According to geomorphologic studies, most part of the area belongs to Quaternary and is formed by alluvial cones with loose and sandy sediments.

During the period 1938-1978, the area crossing the present line registered 166 earthquakes, three of which were of large magnitude.

Gravimetrical Remeasuring of the Mendoza - San Juan - Guayamas Line

Cerrato et al. [1983] remeasured gravity along the Guayamas - San Juan - Mendoza line in 1982. The initial measurement was made in 1949, so that the survey interval was 33 years. Even though they belong to a very close line, the stations occupied between San Juan and Mendoza are not the same as those corresponding to the N 24 line measured by the Instituto Sismologico Zonda, with the exception of the end point.

The difference of +0.05 mGal between San Juan and Mendoza (considered as no change ?) is not significant, whereas in the section, from Caucete to the southwest of Pie de Palo, that was practically destroyed by the earthquake of November 1977, gravity changed for -0.25 mGal in 33 years relative to near stations. Volponi et al. [1982] pointed out that levellings performed in this section (see Figure 2) revealed irregularities originated on the fact that the measuring piers are located on the alluvial bed of the Tulum valley. Deformations due to seismic movements as well as to changes in the discharges of the aquiferous used for irrigation took place in this valley.

It is necessary to point out that when Mendoza was crossed with a transcontinental profile, which starts from Buenos Aires and was remeasured after 30 years [Cerrato, 1979; Figure 4], a gravity difference of -0.66 mGal was found, reaching to its maximum of -1.47 mGal towards the Andean dorsal. Even though conclusions over the relative behaviour among Guatamas, San Juan and Mendoza are maintained in a general frame, the difference of -0.66 mGal as found in Mendoza will meet some questions.

Andean Cortical Movements in South Argentina

In the Andean cross section of the 39°S parallel (see location in Figure 5) a gravity crustal model was calculated with the following values:

crustal density: 2.9 gr/cc,
upper mantle density: 3.3 gr/cc,
crustal normal thickness: 33 km.

The Bouguer anomaly is almost a mirror picture of topography (maximum altitude: approx. 2000 m; maximum Bouguer anomalies: -110 mGal). The gravity model justifies the Bouguer anomalies with a crust which is 43 km thick under the highest altitude [Diez Rodriquez, 1984, unpublished]. If we use the isostatic - floating theory, the regional isostatic anomalies under the Andean belt will be positive (maximum: +35 mGal). Then, the root is not sufficient, or the topography is under compensation. The extensional deformation for the sector has been recognized, so everything makes us to

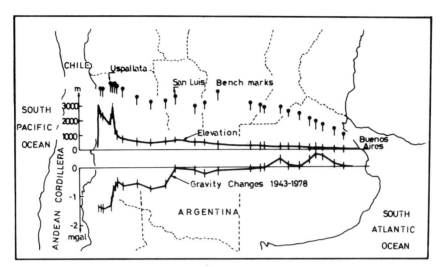

Fig. 4. Gravity changes (1943-1978) along the profile from Buenos Aires to Andean structure [Cerrato, 1979].

suppose that the mountain ranges would be sinking toward a new isostatic equilibrium.

Cerrato [1978] remeasured gravity at 19 stations along a profile, 600 km long close to the 41 S latitude line, after almost 25 years (see location in Figure 5). This profile links San Antonio Oeste with Puyehue path on the Chilean border. The latter area is relatively close to the above analyzed section. Almost all the changes are positive and approach the maximum values of -0.60 to -0.70 mGal, to the west. If 39° S parallel profile exists - unluckily this parallel has not been remeasured - those differences would confirm the descending isostatic process.

Subsidence and High Gravity in the Two Big Atlantic Bonaerense Basins

Geohistoric diagrams were drawn for the two big sedimentary Bonaerense basins of Salado and Colorado (see location in Figure 6). They have 6500 to 7000 m of Cretaceous and Cenozoic sediments, according to drilling data. Curiously, both basins show high gravity over the existing thick sedimentation and active subsidence (recognized as of the observations and tendency inferred from the last portion of the geohistoric diagrams).

In effect, the following maximum Bouguer anomalies are present in the central part of the basins [Introcaso, 1980, 1982]: Salado basin +55 mGal; Colorado basin +31 mGal. And if we correct these values for the effects of the sediments, we have: Salado basin +110 mGal; Colorado basin +120 mGal.

A recent proposition [Introcaso and Ramos, 1984] suggests that the Salado basin is related with the opening of the South Atlantic in an aulacogenical way. The process involves expansion of deep materials from asthenospherical levels which provoke stress on the crust with plastic flow in the lower part and with block faulting and subsidence in the brittle upper part. Then the subsidence would be controlled by cooling and sediment loading.

The neighbouring Colorado basin which, according to seismic data, presents an attenuant crust [Ewing and Ludwig, 1963], would have a similar behavior.

Belousov [1962] mentions that the areas near Buenos Aires and La Plata subside at a velocity of 1 and 10 mm/y, respectively, while in the eastern section of Figure 6 we can see that gravity would have increased in these 35 years. These results seem to confirm the subsidence of the basin edge at a velocity higher than its uoterior filling. Let us remember that the basin has a tendency not only to subside but also to enlarge.

The last section of 10,000 years in historic diagrams points out a subsident tendency (more evident for the Colorado basin).

We conclude that, at the present time, both of the basins present active subsidence. We would be facing a process of not-finished evolution which will minimize the outstanding positive gravimetrical responses, in the geological future.

Analysis of Gravity Measurements

First Measurements

Central Profile (Mendoza - San Juan - Guayamas line) and Buenos Aires (Bs. As.) Province were measured 30-35 years ago by Buenos Aires University. They used the Mott-Smith C-14 gravimeter with temperature stabilizer, and worked in the looper-motion scheme, repeating station readings in such a way as A-B-A-B-C-..., for the best results.

Northern and Southern Profiles were surveyed with a Worden 561 gravimeter with no temperature

Fig. 5. Geographic location of various profiles. Northern Profile (remeasurements of "g" for the section, Resistencia - Cabeza de Buey; will be continued until Socompa); Central Profile (remeasurements of "g" for the section, Buenos Aires - Las Cuevas); Southern Profile (remeasurements of "g" for the section, San Antonio West - P°Puyehue); Transcontinental gravity profile in 39°parallel (without remeasurements of "g"); Transcontinental gravity profile in 32°S (with remeasurement of "g" for the section, San Juan - Guayamas - Chepes. Altitudes have been remeasured between Mendoza - San Juan - Caucete ...).

stabilizer, and in the operation scheme same as above. The route San Juan - Caucete was measured by Rosario University, using a Worden Geodetic instrument with temperature stabilizer.

Remeasurements

Central, Southern and Northern Profiles as well as Bs. As. Province were measured by Bs. As. University, using LaCoste and Romberg gravimeters, G-190 and G-194. The operation scheme was such as A-B-C-D-...Y-Z--Z-Y-...D-C-B-A, this time.

The results were corrected for tidal effects. From comparison with the standard base station, the following calibration factors were obtained by Bs. As. University: Mott-Smith, $F_{MS} = 1.00168$; Worden 51, F_W(Northern Profile) = 0.994807; F_W (Southern Profile) = 0.993728; LaCoste Romberg, LCR G-190, $F_{G190} = 1.000213$; LCR G-194, $F_{G194} = 1.000346$.

The mean square error of a single observation in Northern, Southern and Central Profiles was ±0.02, ±0.02, and ±0.03 mGal, respectively. Gravity tie Buenos Aires - Las Cuevas with LCR G-190 and G-194 showed a very little difference, as Δg (LCR G-190) = 1049.65 mGal, and Δg (LCR G-194) = 1049.66 mGal.

Postseismic Remeasurement

The route San Juan - Caucete ... was surveyed after the earthquake of November 1977 with LaCoste Romberg 145 gravimeter (Instituto Geografico Militar) and then with Sodin 410 gravimeter of no temperature stabilizer (Instituto Sismologico Zonda), in the operation scheme A-B-A-B-C-... Before and after field work, all the gravimeters, except Sandin, were carefully tested at the standard calibration base.

In Prof. Cerrato's opinion, careful operation like this could eliminate systematic errors from observations. Briefly, the gravity difference found in the epicentral area of the 1977 Caucete earthquake is accurate enough, according to Volponi [1982].

Levelling Data

Most of levelling surveys were performed by Instituto Geografico Militar Argentino. So, Mendoza - San Juan - Chepes line, East-West profile along 32°S parallel and East-West profile along 39°S parallel were performed with levelling of high precision (1 mm $\sqrt{S}$ km), using Zeiss III level with parallel plate and invar staves.

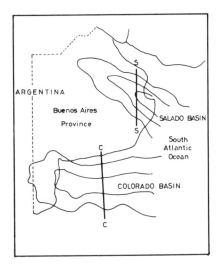

Fig. 6. Salado Basin and Colorado Basin in Buenos Aires geological province.

Relevellings on Mendoza - San Juan and San Juan - Caucete - Chepes were performed by Instituto Sismologico Zonda (after 30 years) using a universal automatic Wild NA 2 level (3mm $\sqrt{S}$ km) and invar staves.

Both, levelling and relevelling were performed according to the first order standard.

Discussion and Conclusions

Volponi et al's [1982] explanations for inverse changes between level and gravity variations, as recorded in Pie de Palo, would correspond to the relative uplifting movement of a rigid block with no significant mass alterations, at least in the upper and medium crust.

The present tendency of relevellings seems to indicate that uplifting continues together with new associate earthquakes. In effect, in the lapse of time between 1938 and 1976, velocity was +0.5 cm/y. Movements associated with Caucete earthquake of November 1977 produced an uplifting higher than 1 m, whereas the present uplift velocity (from 1978 to 1980) is more than 1 0 cm/y.

The results obtained by Cerrato et al. [1983] for Caucete, which was destroyed by the earthquake in 1977, give decrease in gravity relative to near stations. In reference to the relevellings by Volponi et al. [1982], it seem to indicate a direct relationship between changes in gravity and in altitudes. Everything makes us to suppose that significant changes in masses have taken place (an important liquefaction of grounds have recognized in that area).

Previous studies have recently been extended towards the longitudinal line which connects the cities of San Juan and Mendoza. Within a general frame we find apparent puzzling between differences in gravity and differences in altitude, as follows:

Considering that Chepes (la Rioja) is stationary, the difference in altitude in the cities of San JUan and Mendoza would have decreased for 9 and 14.5 mm/y respectively in 44 years while the difference in gravity would not be significant 80.05 mGal/33 y). Nevertheless, if we consider the variations of "g" found over the transcontinental profile, Mendoza city's gravity would have decreased for -0.66 mGal in 35 years. This raises the following questions:

i) If it is true that Chepes remained stable (that is to say, no change in elevation), then we would have to admit that San Juan and Mendoza subsided as reported by Sisterna et al, [1982]. At the same time, if we condsider gravity decrease at Mendoza, as reported by Verrato et al. [1983], then San Juan would also have experienced a decrease of "g," because the difference of "g" remeasurements between both cities is approximately zero. So that significant changes in masses would be present in this longitudinal belt.

ii) Gordillo and Lencinas [1972] studied the Pampean Ranges of San Luis and Colorado, neighbouring to Chepes, and reported compressional movements as well as significant Plio-pleistocenic uplifting of the whole area. In this way we would be far from quaranteeing the stability of Chepes.

In the future, the isostatic behaviour of Los Andes will be analyzed on that section devoid of Quaternary volcanism, where Volponi [1979] reports a strong coupling of plates (plains segment) on the basis of hypocenter distribution. Regarding this area, Baldis et al. [1979] point out that the strong Andean thrusts from W to E oppose with the Pampean thrust from E to W;

On the profile close to the 41°S latitude line, the regional gravity increase seems to indicate the crustal subsidence. On the other hand, the gravimetrical section which ends at Puerto Saavedra (Chile) and is near the previous section, shows excess in gravity under the Andean dorsal.

There is Quaternary volcanism in those latitudes. The Nasca plate is clearly decoupled from the South American plate. Besides, distension has been admitted here. We believe that isostatic forces might be operating to establish a new equilibrium.

Nevertheless, we must be very careful for trying to interpret both profiles on the same basis, as they are separated for approximately 2° in latitude.

Finally, let us say about high gravity and admitted subsidence seen in the Bonaerense or Atlantic area. This area belongs to the Pampean plain, the big sedimentary basins of Salado and Colorado, which is generically related to the opening of the Atlantic Ocean. Besides, at the present time, an increase of "g" in these 35 years has been mentioned at the N-E edge of the Salado basin. This indicates that the subsidence is associated with widening of the basin, where sedimentary filling is insufficient.

On the basis of all the available data, we predict that the process of subsidence will continue in both basins. A study of gravity variations along a transversal section in the Salado basin has been scheduled.

References

Baldis, B., E. Uliarte, and A. Vaca, Análisis estructural de la comarca sísmica de San Juan, Rev. Asoc. Geolog. Arg., 34, 294-310, 1979.

Belousov, V., Problemas Básicos de Geotécnica, Edit. Omega, 854 p., 1962.

Cerrato, A., Remedición de perfiles gravimétricos, comparación de sus resultados con los obtenidos hace 25 ó 30 años, Public. Instituto de Geodesia, Univ. Nac. de Buenos Aires, 13 p., 1979.

Cerrato, A., J. Masciotra, and O. Nuñez, Remedición de la línea gravimétrica Mendoza - San Juan - Guayamas, 1949-1982, Instituto de Geodesia, Univ. Nac. de Buenos Aires, 26 p., 1983.

Diez Rodríquez, A., Perfil gravimetrico Argentino - Chileno, en el paralelo 39°S, Instituto de Física de Rosario (inédito), 1984.

Ewing, M. and W. Ludwig, Geophysical investigations in the Submerged Argentine coastal plains from Bs. As. to Peninsula de Valdez, Bull. Geolog. Soc. Am., 74, 275-292, 1963.

Gordillo, C.E., and A. Lencinas, Sierras Pampeanas de Córdoba y San Luis, Geolog. Reg. Arg., Ac. Nac. de Córdoba, Argentina, 133-159, 1972.

Introcaso, A., and E. Huerta, Perfil gravimétrico transcontinental en el paralelo 32°S, Rev. I.P.G.H., 21, 133-159, 1972.

Introcaso, A., A gravimetric interpretation of the Salado Basin (Argentina), Bollettino di Geofisica Teorica ed. App. 22, 87, 187-200, 1980.

Introcaso, A., Dos modelos que explican la alta gravedad en cuatro cuencas Atlanticas, Salado, Colorado y San Jorge (en Argentina) y Santa Lucia (Uruguay), Instituto de Física de Rosario (IFIR), 8 p., 1982.

Introcaso, A., and V. Ramos, La cuenca del Salado: un modelo de evolución aulacogénica (inédito), 1984.

Sisterna, J., A. Robles, and Munizaga, Renivelación de la línea No 24 San Juan - Mendoza, Instituto Sismológico Zonda (ISZ), San Juan, 31 p., 1982.

Triep, E., Source mechanism of San Juan Province earthquake, Instituto Sismológico Zonda (ISZ), San Juan, 11 p., 1979.

Triep, E., and C. Cardinali, Mecanismos focales en la megafalla Desaguadero-Bermejo y regiones vecinas, Geoacta (en impression), 1982.

Volponi, F., Informe del Simposio Binacional Argentino-EE.UU. sobre el terremoto de Caucete del 23 de noviembre de 1977, Instituto Sismológica Zonda (ISZ), San Juan, Argentina, 44 p., 1979.

Volponi, F., J. Sisterna, and A. Robles, Orogenia: Fuerzas gravitacionales y fuerzas tectónicas, 5° Congreso Lat. de Geologia Argentina, Actas III, 719-730, 1982.

Wollard, G.P., Crustal structure from gravity and seismics measurements, J. Geophys. Res., 64, 1521-1544, 1959.

SUMMARY OF THE MID-TERM REPORT, WORKING GROUP 1, ICL: PROGRESS IN THE FIRST PHASE OF THE ILP

Keichi Kasahara

Earthquake Research Institute, University of Tokyo, Japan[1]

Working Group 1 (WG1) of the Inter-Union Commission on the Lithosphere (ICL) has been active since 1980, providing international and interdisciplinary coordination of research on Recent plate motion and deformation. In addition to exchanging information among WG1 members and other cooperating groups, a series of scientific symposia, most of which were co-sponsored with other groups, were conducted in various parts of the world.

As with other areas of solid-earth science, there has been notable progress in research in this field during the past five years — probably more than was expected when WG1 was established. We are pleased that we could participate, even if indirectly, in this great progress. The International Lithosphere Programme has now completed its initial phase, and this summary of our Mid-Term Report will be linked to our initial objectives and will discuss progress with respect to each objective individually.

On-Going Motion of the Tectonic Plates

The establishment of astrogeodesy, or space geodesy, has provided powerful tools for experimental measurement of on-going plate motion. The pioneering work in the US and European countries has made it possible to operate many stations, fixed and mobile, in various areas of the world.

The first encouraging approach to the goal of direct measurement of plate motion was reported by Anderle and Malyevac (1983) and Christodoulidis and Smith (1983), successively. More recently, the US-Japan cooperative VLBI experiment across the Pacific has provided more immediate evidence for on-going motion of this huge plate (Kondo et al, 1985). Although these results are only preliminary, they support the optimistic view that during the next five years there will be direct experimental proof of motion among two or three major plates.

Progress with space geodesy technology has also had a

---

[1] Present address: Science & Engineering Research Laboratory, Waseda University, Kikui-cho, Shinjuku-ku, Tokyo 162, Japan

Copyright 1987 by the American Geophysical Union.

strong impact on research in other fields such as local tectonics and polar motion studies (see (b) and (f)).

Deformation of the Plates at Their Boundaries

Plate margins occupy only a fraction of the total surface area of the plates, but most plate interaction is concentrated in these narrow zones. Measurement of deformation at plate margins has been especially intensive in several selected areas of high seismic risk. Numerical simulation has been a useful technique for the study of physical mechanisms of subducting plate system (e.g. Hashimoto, 1985), where the rheological property of rocks plays a critical role.

A number of investigations in ocean trenches have extensively improved our knowledge about the mechanical aspect of plate boundaries (e.g. Nakamura, 1986, in this Report). Nakamura's speculation (1983) is notable in this respect. He suggests that the east Japan Sea deformation zone may represent a nascent convergent zone between the North American and Eurasian plates. If this is the case, then the boundary of the two major plates has only recently "jumped" to the above-mentioned line, leaving the former boundary across Hokkaido inactive.

Stress and Strain in the Lithosphere

Technical developments and an increasing demand for precise knowledge about crustal stress/strain conditions have accelerated the acquisition of field data. Such stress information is, in principle, microscopic. It represents only local conditions at the observation site. For a macroscopic view of the crustal stress field, all reliable data must be carefully compiled over a wide area. The stress field map of the North American continent by Zoback et al. (1984), and that of other major plates by the Circum-Pacific Map Project (1985) are examples of this kind.

A major concern is temporal changes of the crustal field, particularly their connection with earthquake processes. The precise monitoring of stress changes in the field and the laboratory testing of loads on large rock specimens, have been actively studied in order to develop earthquake prediction techniques (see Sobolev, 1986, in this Report). A computer simulation of mechanical processes in the crust (e.g. Loo, 1986, in this Report) is another notable advance in this field.

Crustal Movements Associated With Major Earthquakes

Remarkable progress has also been made in this research field in many respects, e.g. field observation, instrumentation, data acquisition and theory. Advanced systems of observation have been constructed for earthquake prediction work and comprise a combination of precise geodetic surveys with a dense network of geophysical instruments. Observations are normally telemetered to data centers for real time monitoring of crustal movements which might be precursory to major earthquakes. Examples of integrated systems exist in California, USA (e.g. National Research Council, 1981) and in Tokai district, Japan (e.g. Suzuki, 1984).

The trenching of active faults (Sieh, 1978) has been widely introduced. Its aim is to study the history of recurring fault activity by excavating soil layers preserving past seismic offsets.

There have been a considerable number of major earthquakes in these five years. Some of them have occurred in unpredicted places (e.g. Central Japan Sea earthquake, 1983), but others were close to predicted locations (e.g. Mexican earthquake, 1985), and almost within a predicted time-window (Adak Is. earthquake, 1986 (Kisslinger et al, 1985)). Optimistically, these examples show some progress in the successful prediction of earthquakes. The Parkfield, Calif., experiment, which started last year, represents a major initiative in earthquake prediction. The time-window set for this work is 1988 + 2 years. This corresponds with the remaining period of the ILP and the experiment is likely to form one of the major topics of this period.

The geodetic application of the GPS (Global Positioning System) is now being tested in several countries. This technique is of great interest to WG1 because of its excellent portability, high precision and long range of distance measurement. It will be a powerful tool for the study of tectonic movements on various scales, in complement with other space-geodetic techniques.

Vertical Movements of the Lithosphere

This problem has been studied extensively in many countries, using various approaches, from geodesy to geology. An on-going effort has accumulated a great amount of data, on vertical movements of the lithosphere, in both seismic and aseismic situations. International geodetic committees, e.g. the Commission on Recent Crustal Movements (CRCM), have been promoting the international linkage for work of this kind. The compilation of vertical land movements in Europe (Vyskocil, 1986, in this Report) represents an example of the result.

Vertical movements of the lithosphere lead to an interest in the anelastic behavior of sub-crustal layers. The Commission on the Quaternary Shorelines and groups in the International Geological Correlation Programme (IGCP) have been coordinating international cooperative studies on the bases of tide-gage and shoreline data (e.g. Pirazzoli, 1985, in this Report) to study the recent movement of oceanic lithosphere.

Polar Motion and Earth Rotation

The geodetic application of space techniques has enabled us to observe these parameters with a precision of miliardseconds (e.g. Ye and Yokoyama, 1986, in this Report), that is an accuracy a hundred times as high as the accuracy of traditional optical methods. This new trend in observations has seriously affected the world's astrogeodesists and led to the proposal of a new organization, the International Earth Rotation Service. This Service, which is planned to start in 1988, will revolutionize this research field and global geodesy by the establishment of a world geodetic reference system.

Data File/Base for the Theoretical Modeling of Internal Processes

The increasing activity as outlined above is rapidly accumulating a great deal of data in various disciplines. Efficient use of these data would be impossible without the aid of modern data management systems. Within individual disciplines this is being addressed. However, for interdisciplinary work, the situation remains undeveloped. The compilation of a geodynamic map by the Circum-Pacific Map Project (1985) is a valuable initiative in this respect.

The above is a brief summary of progress in and around the field of WG1. In general, most areas of research are entering a period of rapid development on the basis of new concepts and technology. We confidently expect significant further developments of the International Lithosphere Programme in the coming years.

Problems for Future Work
(Recommendations)

As seen above, the research with which we are concerned has achieve significant progress in these five years. The coordination of international cooperation, requested of the Group at the beginning of the ILP, has been successful, at least at the preparation level. Through the work of WG1 much has been learned about the problems of coordination. However, the Group notes a number of points which will be important in the next phase of research:

1. Considering the rapid progress in space geodesy, the experimental proof of contemporary plate motion should reach its primary goal by the end of the ILP, at least for two or three major plates. However, further precision will be required to determine, for example:
   (a) variations of their velocity in time and space,
   (b) the motions and deformations of smaller plates,
   (c) expected disagreements between theory and observation.

2. Sooner or later, the simple assumption of a totally rigid plate will have to be abandoned in favour of the concept of a deformable plate. This is especially critical for studying plate boundary problems, the mechanical behavior of micro-plates, and the connection with seismotectonics. As a result of this trend, subjects (a),

(b), (c), (d) and even (e), treated separately in the initial work plan, will become more integrated.

3. The close and seamless linkage of intercontinental geodetic networks (e.g. VLBI) with local systems will be essential for successful progress. New local surveying techniques, such as the geodetic application of GPS, must be developed quickly and applied widely for this purpose.

4. The movements of oceanic plates, both contemporary and in recent geologic time, are poorly known. More effort must be made to solve this problem. The development of submarine instruments is urgently needed together with geomorphic surveys of coastal areas. The anelastic behavior of a plate system will be a major topic in this area.

5. The compilation of interdisciplinary data and the construction of theoretical models of the plate system are needed in order to deepen our knowledge about the dynamics of plates. The coordination of existing data is highly desirable, although it is recognized that there are many technical and administrative difficulties. It may be useful to examine the possibility of a directory for these existing files.

6. In general, we believe that inter-disciplinary cooperation with other groups and committees sharing common interests is extremely useful. The group recommends that relationships should be strengthened to include these colleague groups (see the Executive Summary, for our recommendation to the ICL, 1985).

References

Anderle, R.J. and C.A. Malyevac, Plate motions computed from Doppler satellite observations, presented at IUGG Symposium 2, Hamburg, 1983.

Christodoulidis, D.C. and D.E. Smith, The role of satellite laser ranging through the 1990's, NASA Technical Memorandum 85104, 1-23, 1983.

Circum-Pacific Map Project, Geodynamic Map of the Circum-Pacific Region, Circum-Pacific Council for Energy and Mineral Resources, Am. Assoc. Petrol. Geologists, 1985.

Hashimoto, M., Finite element modeling of the 3-dimensional tectonic flow and stress field beneath the Kyushu Island, Japan, J. Phys. Earth, 33, 191-226, 1985.

Inter-Union Commission on the Lithosphere, Dynamics and Evolution of the Lithosphere — The Framework for Earth Resources and the Reduction of Hazards, ICL Report No. 1, 62 pp., 1981.

Kisslinger, C., C. McDonald, and J.R. Bowman, Precursory time-space patterns of seismicity and their relation to fault processes in the central Aleutian Islands seismic zone, Symposium No. 1, IASPEI General Assembly, Tokyo, August, 1985.

Kondo, T., K. Heki and t. Yakahashi, Pacific plate motion detected by the experiments conducted in 1984-85, J. Radio Res. Lab., 1986 (to be published).

Loo, Huan-Yen, Three-dimensional numerical analysis of continental marginal basin deformation related to large earthquake development, Mid-Term Report, WG1, ICL, 1986 (in press).

Nakamura, K., Possible nascent trench along the eastern Japan Sea as the convergent boundary between Eurasian and North American plates, Bull, Earthq. Res. Inst., Univ. Tokyo, 58, 711-722, 1983.

Nakamura, K., Trench depth and relative motion between overriding plates, Mid-Term Report, WG1, ICL, 1986 (in press).

National rEsearch Council, Geodetic Monitoring of Tectonic Deformation - Toward a Strategy, National Academy Press, pp. 109, 1981.

Pirazzoli, P.A., Lithospheric deformation as deduced from former shorelines: a review, Mid-Term Report, WG1, ICL, 1986 (in press).

Sieh, K.E., Prehistorical large earthquakes produced by slip on the San Andreas fault at Pallet Creek, California, J. Geophys. Res., 83, 3907-3939, 1978.

Sobolev, G.A., Stress and strain measurements in the USSR by geophysical and tectonophysical methods, Mid-Term Report, WG1, ICL, 1986 (in press).

Suzuki, Z., The fifth 5-year program for earthquake prediction in Japan, Japan-China Symposium on earthquake prediction, Tokyo, December, 1984.

Vyskocil, P., Recent crustal movements on the European continent, Mid-Term Report, WG1, ICL, 1986, (in press).

Ye, Shu-Hua and K. Yokoyama, Interaction between the earth's rotation and the plate motion, Mid-Term Report, WG1, ICL, 1986 (in press).

Zoback, M.L., M.D. Zoback, and M.E. Schiltz, Index of stress data for the North American and parts of the Pacific plate, U.S. Geological Survey Open File Report 84-157, 1984.